Resource Efficiency Complexity and the Commons

T0227426

The efficient use of natural resources is key to a sustainable economy, and yet the complexities of the physical aspects of resource efficiency are poorly understood. In this challenging book, the author proposes a major advance in our understanding of this topic by analysing resource efficiency and efficiency gains from the perspective of common pool resources, applying this idea particularly to water resources and its use in irrigated agriculture.

The author proposes a novel concept of 'the paracommons', through which the savings of increased resource efficiency can be viewed. In effect, he asks: who gets the material gain of an efficiency gain? By reusing, economising and avoiding losses, wastes and wastages, freed up resources are available for further use by four 'destinations'; the same user, parties directly connected to that user, the wider economy, or returned to the common pool. The paracommons is thus a commons of – and competition for – resources salvaged by changes to the efficiency of natural resource systems. The idea can be applied to a range of resources such as water, energy, forests and high-seas fisheries.

Five issues are explored: the complexity of resource use efficiency; the uncertainty of efficiency interventions and outcomes; destinations of freed up losses, wastes and wastages; implications for resource conservation; and the interconnectedness of users and systems brought about by efficiency changes. The book shows how these ideas put efficiency on a par with other dimensions of resource governance and sustainability such as equity, justice, resilience and access.

Bruce Lankford is Professor of Water and Irrigation Policy at the School of International Development, University of East Anglia, UK. He has 30 years' experience in the field of land and water development, mainly in Sub-Saharan Africa, and is a co-founder of the UEA Water Security Research Centre.

'Throughout the world, water managers and users face numerous concerns that collectively define a water crisis: growing competition for supplies of water, the need to grow more food with limited water, environmental pollution and degradation, and dealing with floods and droughts. Gains in water use efficiency are frequently hailed as the number one solution. However, others have pointed out that misuse and misunderstanding around efficiency could lead to counterproductive action – if farmers consume more scarce water supplies by becoming more efficient they may leave less for others – or that efficiency gains could result in less return flows back to the environment. Ultimately a more complete picture of water efficiency is required. Bruce Lankford addresses these critical issues head on, unpacking the term efficiency in water resource management to understand the winners and losers, costs and benefits of our water actions. This book is essential for water practitioners and researchers, and others involved in natural resource management, if we really want progress in water management.'

David Molden, Director General, *ICIMOD, Nepal*

'Bruce Lankford has written an exciting and innovative book. *Resource Efficiency Complexity and the Commons* develops the novel concept of 'the paracommons' through which savings of increased resource use efficiency can be viewed. It throws a new and bright light on how these efficiency gains in resource use may be made and transmitted and who benefits from them. Policy makers, scientific advisors and members of select committees should read this book, as well as graduates, research students and academic faculty.'

Piers Blaikie, Emeritus Professor, *University of East Anglia*

Resource Efficiency Complexity and the Commons

The paracommons and paradoxes of natural resource losses, wastes and wastages

Bruce Lankford

Routledge
Taylor & Francis Group
LONDON AND NEW YORK

earthscan
from Routledge

First published 2013
by Routledge

2 Park Square, Milton Park, Abingdon, Oxfordshire OX14 4RN
711 Third Avenue, New York, NY 10017

Routledge is an imprint of the Taylor & Francis Group, an informa business

First issued in paperback 2017

British Library Cataloguing in Publication Data
A catalogue record for this book is available from the British Library

Library of Congress Cataloging-in-Publication Data
Lankford, Bruce A.
  Resource efficiency complexity and the commons : the
  paracommons and paradoxes of natural resource losses, wastes and
  wastages / Bruce Lankford.
      pages cm
  Includes bibliographical references and index.
  1. Waste minimization. I. Title.
  TD793.9.L355 2013
  333.2--dc23   2013007389

ISBN: 978-0-415-82846-8 (hbk)
ISBN: 978-1-138-57474-8 (pbk)

Typeset in Bembo
by HWA Text and Data Management, London

Contents

Figures

Tables

Boxes

Preface

The paracommons is an idea describing the competition over future resources 'freed up' by efficiency gains. The paracommons is interested in the consequences of today's wastage and waste being reduced in the future. The prefix 'para' indicates that the paracommons sits alongside 'the commons' (e.g. water in a river or fish/fisheries in the sea). The paracommons is interested in how changes are made to resource use efficiency and the complexity of verifiying this endeavour over time and space. By using this term, and the word 'paragains', I hope to capture the sense that resource efficiency science does not fall into a binary that says either a) resources are always freed up by efficiency gains or b) no resources are freed up by efficiency gains. These two camps have their respective protagonists. In addition, the paracommons asks who gets (or owns) the material gain of an efficiency gain. The message of the paracommons is that resources found by savings and efficiency gains exist but their size, character and destination are 'liminal' pending the resolution of many different context-specific factors such as technology, scale and measurement.

The idea of a liminal paracommons is not immediately apparent. It essentially says that 'savings' of the inefficient part of resource use is complex, provisional and a matter of common-pool competition. The questions at the heart of the book are; 'who gets the gain from an efficiency gain, what is an efficiency gain and what mediates gains? The idea can also be applied to the changing emphases in ecosystem services that dictate the nature of 'losses, wastes and wastages'. Yet from recent experience, I don't deny that first-time readers might struggle to understand what I am trying to say – and I wonder whether this is a reflection of under-developed science of resource use efficiency. Too frequently I see efficiency being used in a simplistic form, not mentioned, being wished away altogether, or utilised in a modernistic fashion to impart validity to other ideas such as green growth.

I started work on resource efficiency complexity during the middle of 2011 while writing on the science of irrigation efficiency (Lankford, 2012a, 2012b). Readers may see the connections between that paper and this book; asking who gets and owns the waste fraction of a recoverable resource undergoing an efficiency change in a resource-scarce world. Furthermore this book adds to the debate in the journal 'Water International' (JWI) during 2011–12 (Frederiksen and Allen, 2011; Gleick *et al.*, 2011; Frederiksen *et al.*, 2012). However, without comprehensive field research and empirical data – which to my mind does not exist – this book is unable to answer the question of 'who gets the gain of an efficiency gain?' Instead, very much in keeping with current literature (e.g. the JWI papers), a series of models and worked examples are employed. While in the case of irrigation, gains tend to be appropriated by irrigators themselves, this book shows that other destinations and parties have claims on those resource gains. It is the common pool nature and materiality of efficiency gains that I wish to highlight.

This book is trans-topic in exploring the efficiency of resource use; it uses water, forests, energy, carbon and ecosystem services. But I am not sure the book is trans-disciplinary – while it hints at the economic and social treatment of efficiency, I primarily wish to tackle efficiency from a physical, natural science point of view. I don't deny that any scholars working deeply within one particular field of irrigation efficiency, energy efficiency or ecosystem services will feel dissatisfied with the treatment of their particular field. Or that economists, sociologists, engineers and ecologists will see an insufficient explication of their disciplinary treatment of efficiency. Yet for other scholars working on environmental sustainability there may be too much depth and detail regarding efficiency. I have chosen a path through various literatures and ideas in order to sketch out the meaning of the paracommons.

To communicate these ideas, I have used tables and diagrams, cross-referencing and repetition in places. Nevertheless, communicating the concepts of the paracommons and liminality has presented difficulties. For about 12 months I called the idea 'the liminal commons' but this seemed not to have much communicative potency with colleagues. At one point I came close to using the term 'the paradox commons'. However, I chose the term 'paracommons' because although paradox captures unpredictabilities, there is more here than paradoxical outcomes. The Greek prefix 'para' also signifies 'alongside', describing the connective distributive outcomes part of the theory and of the idea of a pending competition over the 'paragains' freed up by efficiency gains. 'Para' can also mean an abstraction of the idea – this is entirely appropriate given that the paracommons is abstract and sits alongside the 'principal' commons.

A frequent response I get when explaining the idea of the paracommons is 'why and what is liminality?' (It is a well-known concept in the social sciences.) The book explains several reasons for choosing liminality as a defining property of the paracommons. Another notion was to revisit the 'inverse commons' – but as I identify in the book, this metaphor has been adopted to express the notion of greater good coming from greater competition and consumption. Neither am I interested in marginality or the marginal commons – although there is some utility in the word 'marginal' in this regard.

Acknowledgements

On this 15-month journey, I'd like to acknowledge Sian Sullivan for immediately recognising what I was trying to say in an earlier version. I drew inspiration from other contributors to the 2012 special issue in the journal *Agricultural Water Management* and their reflections on the complications of irrigation efficiency (to which one can add the energy literature on the Jevons Paradox). Thanks also to Chris Dean (Australia) for his stimulating email correspondence during 2011 about the performance and accounting of attempts to sequester carbon. My thanks to UEA for the sabbatical and to John Mcdonagh for making comments on an earlier version and to Naho Mirumachi, Mark Zeitoun and Tony Allan who at a meal one evening in Norwich in May 2012 urged me to find another term to replace 'the liminal commons'. It was a few days after that I decided to use 'the paracommons'. Discussions with, and work on efficiency by, students at UEA must be recognised: Virginia Hooper, Machibya Magayane, Ruth Makoff and Margaret Bounds. I am very grateful to Tim Hardwick at Earthscan for his reception of the book proposal, to the copy editing team at Earthscan (thank you Holly and Abi), and to the anonymous referees used by Tim and Earthscan whose penetrating comments helped me to improve the manuscript.

Acronyms and abbreviations

CAS	complex adaptive systems
CIE	classical irrigation efficiency
CO_2	carbon dioxide
CP	common pool
CPR	common pool resources (some literatures use 'common property resources')
EIE	effective irrigation efficiency
ES	ecosystem services
ICID	International Commission on Irrigation and Drainage
IFC	International Finance Corporation
IWMI	International Water Management Institute
IWRM	integrated water resources management
JWI	The journal *Water International*
LCA	life cycle assessment
O&M	operation and maintenance (implies management as well)
PES	payments for ecosystem services
PI	Pacific Institute
Pgn	paragain
PWP	permanent wilting point
RAM	readily available moisture
REC	resource efficiency complexity
REDD+	Reducing Emissions from Deforestation and Forest Degradation
SES	social-ecological systems or implicitly 'technological social-ecological systems'
UNEP	United Nations Environment Programme
UNFCCC	United Nations Framework Convention on Climate Change
WE	wider economy

Chapter 1

A preliminary explanation of the paracommons

This book views natural resources and the commons from the point of view of losses, wastes and wastages. The book argues that in a scarce world, competition for losses, wastes and wastages increases. These previously dismissed, unwanted and unavailable (by definition) resources are increasingly scrutinised and sought out. Because losses, wastes and wastages are expressed by resource efficiency, the book explores efficiency and productivity. Thus the book views efficiency gains as a common pool resource expressed in the question 'who gets the material gain of an efficiency gain'? (See Box 1.) The 'who gets the gain' in the first part of the question refers to a material gain as a physical resource, while the 'efficiency gain' in the second part of the question is a performance improvement in the efficiency of a natural resource system. For example, when the physical efficiency of an irrigation system goes from 54 per cent to 59 per cent (this is the performance gain), it means that the irrigation system is now consuming five units more (it 'got' the material gain). However, those five units could have gone to someone or something else. The paracommons is about the competition for the five units 'freed up' by the efficiency improvement. This seemingly innocuous example hides many paradoxes requiring mental agility in the first place and considerable analysis in the second place.

Therefore the paracommons can be seen as a commons of the material gains from efficiency improvements, or put another way; the competition for the inefficient part of resource use. To distinguish competition for wastes and wastages, the prefix 'para' has been added to 'the commons'. Furthermore, the paracommons places efficiency-centred endeavours 'in limbo' problematically located on the boundaries between current efficiency, future intended efficiency, the design of interventions to raise efficiency and final productive, depletive and distributional outcomes. The book explains this framework and contrasts the paracommons with the commons using examples from irrigation and river basins, and carbon, forests and ecosystem services. It is proposed that the paracommons modifies the principle of subtractability (that resources subtracted in one place are not available elsewhere) requiring new thinking on property rights. The Jevons Paradox, when raised energy efficiencies paradoxically fail to reduce aggregate consumption, is one expression of the paracommons.

Box I Freeing up saved resources for other users – three examples

In recent years, the Australian Government has introduced 'water-saving infrastructure projects' for irrigators to 'save' water and deliver water to the river system, mainly through on-farm irrigation efficiency projects. Gains from savings are shared between the irrigators and the river/ environment.[1]

Similarly in the United States, reviews of prior appropriation legislation has meant that efficiency gains associated with irrigators and their water rights can now be debated as in this text from the start of Norris (2011) paper 'the United States Supreme Court's recent decision in Montana v. Wyoming brings to the forefront one of the most complicated and contested facets of irrigation efficiency: who owns the rights to the conserved water?' [2] (In this Yellowstone River case, downstream Montana filed a complaint against Wyoming because the latter's irrigators used up the conserved water created by a technology switch from flood irrigation to sprinkler which reduced drainage water that previously reached Montana.)

Staying in North America – at the transboundary scale, Mexico and the United States examine the Colorado Compact to share water savings made from the lining of irrigation canals and upgrading of irrigation methods in Mexican irrigated agriculture. This agreement is known as Minute 319.[3,4]

Many questions apply. What are the starting conditions? Are selected technologies appropriate? Will these gains really go to those intended and what influences these outcomes?

1.1   The paracommons – step by step

Using the example of irrigation, Figures 1 to 3 introduce how the paracommons arises and how it relates to the commons.[5] All three figures provide a connected sequence and should be read together. Figure 1 begins with a common pool in the upper left-hand corner. In this example it is a body of freshwater in a dam, aquifer or stream. Moving to the next part of Figure 1 (upper right), we see four irrigators or farmers competing over this common pool of water, facing competition and rivalry. In the bottom left of Figure 1, the farmers' water demands are shown as four segments (A, B, C, D) of a water allocation pie, leaving some remaining in the common pool. Moving to the bottom right of Figure 1, part of these irrigation abstractions have efficiency losses. These losses arise, for example, via evaporation of water from bare soil instead of via useful crop transpiration. The four farmers each have four different efficiency levels and thus varying sizes of the 'waste/wastage' fraction of their pie segments.

Moving to Figure 2, the first part in the upper left imports the final bottom-right 'pie' from Figure 1 to continue the story. The net demands (the beneficial

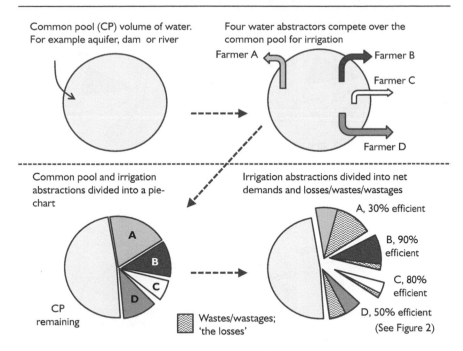

Common pool (CP) volume of water. For example aquifer, dam or river

Four water abstractors compete over the common pool for irrigation

Farmer A

Farmer B

Farmer C

Farmer D

Common pool and irrigation abstractions divided into a pie-chart

Irrigation abstractions divided into net demands and losses/wastes/wastages

A

B

C

D

CP remaining

Wastes/wastages; 'the losses'

A, 30% efficient

B, 90% efficient

C, 80% efficient

D, 50% efficient (See Figure 2)

Figure 1 From the commons to abstraction and inefficient fractions

crop transpiration part) and the inefficient fractions are teased out into two separate diagrams – upper right and lower left of Figure 2 respectively. Finally, by collating the wastes and wastages together as one combined segment, we can show in the bottom right of Figure 2, how 'the paracommons' begins to arise, showing that it sits alongside the commons (or within depending on how you view it).

However, to fully understand the paracommons is to recognise that current 'losses' (or assumptions of losses) frame a wish for a future more efficient system, and that if this higher efficiency is achieved, then these losses are potentially but not inevitably 'freed up' and available to competition. Figure 3 introduces the final crucial part of how a future 'resource efficiency gain' might deliver a reduction in losses and a consequent material gain. In other words, a 'new' resource (termed here a 'paragain') has potentially been 'found' by reducing losses. It is this material paragain that sets up the question 'Who gets the gain of an efficiency gain?' However, so far, everything appears quite linear and predictable; that an efficiency gain can be achieved, and that this frees up a material new resource which is then competed over.

On the contrary, this relationship between efficiency gain, material gain and 'new resource' is not linear, easily achievable and predictable. The translation of a change in efficiency ratio into a new resource is highly complex and mediated by many factors. It is for this reason that I use the term 'resource efficiency

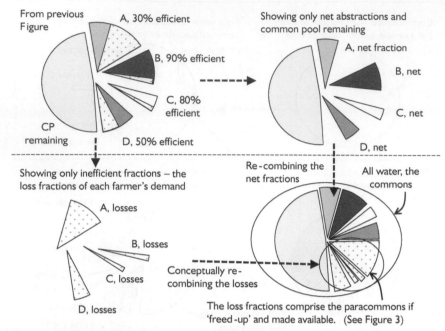

Figure 2 Contrasting the paracommons with the commons

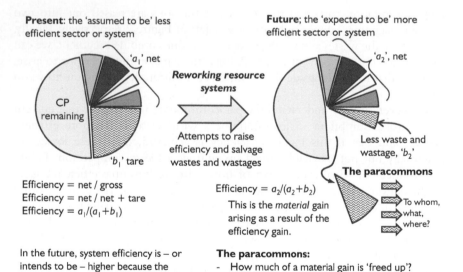

Figure 3 Arriving at a 'paragain' by efficiency improvements

complexity' and argue that new salvaged resources are 'pending' (hence the terms the liminal paracommons and paragains) and behave paradoxically subject to various factors. As explained in the next chapter, and in Figure 11, the story continues with what happens and doesn't happen to this 'new' resource.

1.2 A brief explanation of the word 'paracommons'

Adding the prefix 'para' to the word 'commons' gives the word 'paracommons'. 'Para' signals 'alteration and contrariness' (as in paradox) associated with the unpredictability of the paracommons where the control of losses is problematic and the destination of gains from efficiency gains cannot be anticipated. 'Para' also has a meaning of 'being alongside'. The paracommons puts competition for an efficiency gain alongside competition for resources in their natural capital state. 'Para' denotes another type of 'alongside' (like the word 'parallel') arguing that neighbouring users are physically connected by efficiency and the ensuing destinations for salvaged losses.

> ### Box 2 Etymology of the prefix 'para'
>
> *Para*: Etymology: <ancient Greek παρα- 'by the side of, beside', hence 'alongside of, by, past, beyond', etc., cognate with fore adv. and prep. Also παρά has the same senses as the preposition, along with such cognate adverbial ones as 'to one side, aside, amiss, faulty, irregular, disordered, improper, wrong'; it also expresses subsidiary relation, alteration, comparison, etc. Forming miscellaneous terms in the sense of 'analogous or parallel to, but separate from or going beyond, what is denoted by the root word' (OED, 2012).

1.3 The paracommons and the discarded apple core

Figure 4 reveals aspects of the paracommons by looking at the fate of an apple and its 'to-be-discarded' apple core which ordinarily goes to the city garbage where it might be picked over by people whose lives depended on households throwing away food and goods. For both the apple on a kitchen table and the apple core in the city dump there are two commons. The household members compete over the apple and the city dump 'harvesters' compete over who eats the apple core. However these two commons are separate from each other. What turns this situation into the paracommons are other options that might or might not transpire depending on shifts in efficiency and the core's destination. The household deciding to eat more food (case B in Figure 4) or recycle the apple in garden compost (case C) deprives the waste pickers at the garbage dump of their sustenance. The waste picker fears the outcomes of a household drive to be more efficient and 'green'. Alternatively, a waste furnace (case D) needing waste to generate electricity deprives the householder, the waste picker and the compost of 'their' core.[6] Aspects of the paracommons are:

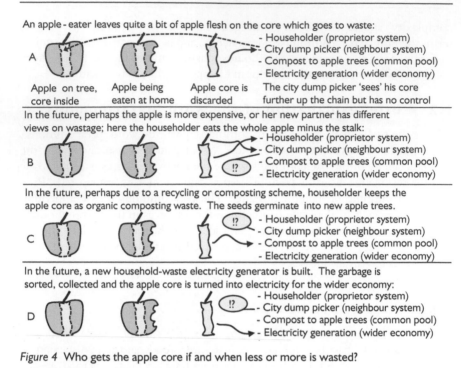

An apple-eater leaves quite a bit of apple flesh on the core which goes to waste:

A   Apple on tree,   Apple being   Apple core is   The city dump picker 'sees' his core
     core inside   eaten at home   discarded   further up the chain but has no control

- Householder (proprietor system)
- City dump picker (neighbour system)
- Compost to apple trees (common pool)
- Electricity generation (wider economy)

In the future, perhaps the apple is more expensive, or her new partner has different views on wastage; here the householder eats the whole apple minus the stalk:

B

- Householder (proprietor system)
- City dump picker (neighbour system)
- Compost to apple trees (common pool)
- Electricity generation (wider economy)

In the future, perhaps due to a recycling or composting scheme, householder keeps the apple core as organic composting waste. The seeds germinate into new apple trees.

C

- Householder (proprietor system)
- City dump picker (neighbour system)
- Compost to apple trees (common pool)
- Electricity generation (wider economy)

In the future, a new household-waste electricity generator is built. The garbage is sorted, collected and the apple core is turned into electricity for the wider economy:

D

- Householder (proprietor system)
- City dump picker (neighbour system)
- Compost to apple trees (common pool)
- Electricity generation (wider economy)

Figure 4 Who gets the apple core if and when less or more is wasted?

- Competition over the apple core takes place between four types of groups; householder, garbage picker, furnace operator and those who speak on behalf of apple tree populations.
- In the paracommons, these four parties are termed 'destinations'. In the same order they are: the proprietor, the neighbour, the wider economy and the common pool (or environment).
- The destination of the core depends on switches and changes in interests and technology affecting the sequencing and cascading of the apple core through the household then onto the city dump, furnace or compost.
- These four destinations cannot easily communicate with each other over who gets the core as do rivalrous commons harvesters inside the house and at the city dump.
- The common pool nature of the core in the paracommons is related to perceptions about today's waste alongside 'savings' to be made in the future. There are spatial and temporal transitions involved.

1.4 The paracommons, households and hospital wards

Introducing the paracommons, some parallels with efficiency gains in two other systems (car fuel budgets and hospital ward/care systems) are sketched out

Table I Efficiency savings within hospital wards

Unit efficiency and its measure	Four destinations; who gets the gain of the efficiency gain	
One hospital ward is the proprietor system in this example The measure is 'time taken per patient per bed' (the bed occupation rate).	One ward makes efficiency gains. It is now more efficient	Costs for this ward drop with possible reduction in number of beds. Can now await an outbreak in disease or bank the savings.
	A neighbouring ward in the ward corridor	A less efficient neighbouring ward takes advantage and sends patients to the efficient ward. Perhaps they have ties and agreements over patient swaps.
	Whole hospital	The hospital gains from the one efficient ward but in a less tangible, less direct way. The ward's allegiance to its immediate neighbour is not so strong when it comes to the whole hospital.
	The common pool ('the public') or environment (energy costs)	Ideally, the more efficient ward, by treating more patients, puts healthier patients back into the population than are generated by ill-health. Or alternatively, lower running costs enable the hospital to reduce its energy bills.

(Table 1 and Box 3). In the hospital ward the efficiency question is how to save resources and beds (measured by bed occupation rate) by the use of improved medical procedures. Essentially the paracommons argues that savings made at the unit level (household or hospital ward) are difficult to ascertain and measure at the system or sector level (neighbourhood or town, hospital or health sector).

And even if they can be ascertained, savings are not easy to 'send' transparently to a particular destination and be 'banked' as aggregate verified savings. The paracommons argues there are four destinations. Using the hospital ward as the example, 'savings' could stay with the individual ward (now having some empty beds or fewer beds), or move to immediate neighbours (two wards in the same corridor might benefit because an inefficient one can send its patients to a neighbouring efficient ward), or move to the wider system (whole hospital treating more patients because of the one ward is being more efficient) or move to the whole population (now healthier).

Instead savings move in complicated ways. Gains are subsumed and savings 'seep' away. The likely outcome is that the hospital puts new emphases on other forms of patient interaction or exports patient care to their homes with the consequent impacts on patient health, or the efficient ward feels 'put upon' as it can never compensate for the rest of the hospital's less efficient ways. The least likely outcome is that the whole hospital (and by extension the whole health sector) genuinely becomes more efficient measured by, for example, healthier populations with a lower health care bill and reduced environmental impact.

Box 3 Who got the gain of an efficiency gain?

Question: what did you do in the last few days that 'saved' something that ordinarily you would not do?

Let's assume the answer is that you cycled to work instead of taking a large family car or that you drove more slowly. Now assume that by cycling or driving more economically you did not consume 1.5 litres of petrol over one week that you would have done.

So, where did the 1.5 litres of savings go – in other words, who got the material gain? The paracommons conceives of four answers:

- You kept the 1.5 litres in your fuel tank and used them at the weekend – in other words you, the proprietor, used them to drive more miles later on.
- You kept 1.5 litres in your fuel tank, but let your daughter syphon it out and use the 1.5 litres in her own car. Your efficiency gain helped another user immediately connected to you. You would not allow a stranger to take fuel from your car but you might see strong connections via the budget of your home and its family members.
- You didn't put 1.5 litres in your car's fuel tank, so it stayed in the fuel bowser at the filling station. (Or you filled up less frequently.) The 1.5 litres were consumed by other drivers whose identity you do not know. Your efficiency gain passed to the wider motoring public (or would have done if the filling station's regular delivery of fuel was delayed by a few hours).
- You didn't put 1.5 litres in the tank as you didn't use it. You genuinely reduced your fuel consumption as a long-term practice. By tracing the fuel savings all the way back to petroleum deposits, the 1.5 litres stayed in the ground, thereby not contributing to global emissions. You helped the environment.

The most likely outcome is the first; that you used the fuel saving another time. The least likely outcome is the last, that you absolutely ensured it did not contribute to global emissions and no-one else used it. Another possible outcome is that it went to all four options in various unknown quantities.

Many complicating factors mean that the one efficient ward's heroic efforts are lost to its neighbours and to its encasing system (the hospital).

Nevertheless, unit-level efficiencies have taken place and have shaped efficiencies and behaviours of the wider systems in which they sit – the problem is these interactions and savings are difficult to trace. Moreover, there is one important difference between savings in the hospital ward, the household fuel budget and savings in irrigation systems. In the ward and house, savings manifest themselves primarily through the owners' economic interests (to save money) blurring the computation of savings because of the changing marginal price and value of a litre of fuel and a bed in a ward. In irrigation systems, while saving money is important, there is a very rich debate on the saving of the physical resource water. It is this physical materiality that makes the commons of losses in irrigation that much more 'competitive'. Thus the competition part of the irrigation paracommons is over saved water, while in households and hospital wards the mix of savings includes energy, time, finance and technology. Nevertheless, in all of these systems the question of who gains from an efficiency gain applies. For example, the house-owner has the choice of transferring her reduced fuel bills to a much less environmentally impactful life, either spending the money on greener goods or by asking for a lower salary which in turn could reduce her employer's economic costs.

1.5 The paracommons – seven clarifications

In this introductory section, I briefly counter seven possible misunderstandings about the paracommons and how the idea is distinct from similar concepts.

1.5.1 The paracommons and the commons

I distinguish 'the commons' framing of competition for natural resources such as rivers, aquifers and forests from 'the paracommons' framing of competition over freed up resources obtained by changing the efficiency of usage. Contrasting the 'commons' with the 'paracommons' is a useful dialectic, although they merge into each other in complex ways. Thus the commons is chiefly about resources withdrawn from natural capital stocks and flows and the paracommons is about resources 'found' by reworking the efficiency of those already withdrawn resources in socio-ecological systems. Therefore the purpose of the paracommons is to discuss physical resource efficiency rather than to discuss the idea of the commons (Wagner, 2012) although later in the book some distinctions are drawn up. In this framing, the term 'common pool nature' covers competition over natural resources and also the competition over salvaged resources freed up during efficiency changes.

1.5.2 Paracommons complexity and commons complexity

Comments on earlier drafts by referees mistook the paracommons as restating the complexity of the commons – a topic that is increasingly emerging within the commons literature (Berge and Van Laerhoven, 2011). Associated with commons complexity, it is possible that readers might think the paracommons addresses economic efficiency, in other words how do attempts to govern the commons result in effective outcomes (for example, Ollivier, 2012). Both are not the case; the paracommons framing of complexity pertains to the technical physical efficiency of natural resource systems and to complexities associated with losses, wastes and wastages. By characterising efficiency as complex, I have attempted to identify a range of factors that shape complexity. A holistic treatment of complexity is also seen in the commons literature – but this does not make the two types of complexity the same.[7]

1.5.3 The paracommons, energy efficiency and the Jevons Paradox

Parts of the Jevons Paradox regarding energy efficiency are found within the paracommons, for example, the ratio of outputs to inputs as the defining of 'efficiency' and the paradox that greater efficiency can lead to a 'rebound' of greater consumption (Polimeni *et al.*, 2008). However, a more complete expression of the paracommons, exemplified by irrigation, deals with the changing and nested material pathways of 'losses' flowing to different pathways destinations when the system undergoes efficiency changes. Furthermore the 'paradox of rebound' occurs for different reasons; the Jevons rebound takes place because of the changing economics of production and consumption driven by efficiency. An efficiency rebound of water higher consumption in river basins (Ward and Pulido-Velázquez, 2008; Crase and O'Keefe, 2009) takes place (it is argued) because of the switch from previously recovered drainage water to consumptive evapotranspiration. Prices do not need to be involved in this case. As such, the paracommons questions the nature of the paradox between efficiency and rising consumption, arguing that one can have 'reducing resource use per unit of economic activity' covered by recent thinking around decoupling (UNEP, 2012).

1.5.4 The paracommons, carbon leakage and the 'green paradox'

In attempting to drive down carbon dioxide emissions through a combination of economic pricing and technical efficiency, scholars have noted that global carbon emissions have continued to rise. This phenomenon has been explored via carbon leakage and the green paradox (van der Ploeg and Withagen, 2012; Eichner and Pethig, 2011; Sinn, 2012). There are aspects of the paracommons found in the green paradox literature, for example, the paradox of efficiency policy that drives up consumption and the lag or limbo between policy intention and policy outcome. However, there are distinctions, one being the framing of common pool competition over resources freed up by technical/material efficiency savings. As with the Jevons Paradox, this distinction relates to the materiality of the paracommons (e.g. water) where the market and prices need not be invoked although clearly these influences exist.

1.5.5 The paracommons, waste recycling and industrial ecology

Recycling of wastes and wastages is found in urban, industrial or agricultural systems (Al Sabbagh *et al.*, 2012). Recycling is also the topic of common pool studies of waste materials ejected by urban populations (Gutberlet, 2008). Recycling is also found in the study of industrial ecology (Bourg and Erkman, 2003); when, for example, one factory's wastes (or public waste) become the inputs for a neighbouring factory (such as chemical effluent, lead in batteries or rare metals in cell phones). Consumption and disposal of goods and wastes are also found in studies of life cycle assessment (LCA) (UNEP, 2012).

The paracommons observes that efficiency is altered in more ways than the recycling of wastes. Furthermore the paracommons examines how users and systems become interconnected by losses, wastes and inefficiencies even prior to recycling. While industrial ecology systems describe a relatively static relationship in the 'here and now', the paracommons is predicated on a highly dynamic context and complex feedback loops, and arises out of a resource gain 'freed-up' by an efficiency improvement. This introduces the idea of a current system and a future system (both states poorly understood and often not measured or monitored). This complexity within a time shift introduces 'systems in transition' and the idea of liminality.

At this point in its development, the paracommons is not an expression of the complexity of the analyses found in LCA – which traces the consumption of natural capital throughout the production, use and disposal of goods and services. Rather, the paracommons is principally about the competition for resources delivered by gains in efficiency and is more particularly interested in the withdrawal and conversion of resources rather than the disposal of goods.

1.5.6 The paracommons and wastewater recycling

Similar to the previous section, I take wastewater recycling, the reuse of a town's effluent for crop irrigation (Qadir *et al.*, 2010) as a version of industrial ecology. The effluent or drainage water is the 'given commons' over which irrigators vie. Through a paracommons lens, wastewater irrigation would be seen in a different light; the paracommons is interested in changing relationships caused by the town altering its efficiency with consequences for its residents and downstream irrigators.

1.5.7 The paracommons and natural resource wastes – industrialising ecology

If industrial ecology covers the web of interactions between industrial units, then an increasing 'industrialisation' (reinventing of nature: Banerjee, 2003; ecological modernisation: Bailey *et al.*, 2011) of the natural world arises from society's rising interest in the reuse of previously discarded resource wastes. Livestock excreta used as biogas and fish waste used for animal meal exemplify the shifting patterns of consumption and recycling in response to supply, demand and ingenuity. Whether or not these represent the paracommons depends on particular factors of complexity, scale and the nature and pathways of inputs, outputs, wastes and wastages.

However, the book concerns itself not with recycling of biological wastes but from more substantial interests regarding the boosting and reduction of different ecosystem services. And it is from this observation that I argue a particular type of paracommons arises from industrialising of ecology and the commons. As explained in Chapters 4 and 5, an increasing concern for the performance of ecosystem services recast by carbon/carbon dioxide provides a version of the paracommons. Efficiency and productivity become more significant in this more evaluative, calculative and objectified environment.

1.6 Terms and definitions

The purpose of this book is to provide a substantive exposition of the idea of the paracommons arising from efficiency-driven changes and complexity in large technological social-ecological systems. It does not aim to provide chapter-and-verse definitions of the myriad terms employed in the study of efficiency and productivity throughout different sectors and literatures. Table 2 offers some reflections on evolving terms and ideas. This framework coins 'only' four new terms: paracommons, paragains, parasystems and parageoplasia. It also employs a term known within the social sciences, 'liminality'.

Table 2 Reflections of some terms employed – not in alphabetical order

Salvage (forestalled, released, recovered, avoided, offset and transferred)	The words 'salvage', 'salvageable' and 'salvaged' cover the five ways that losses associated with a resource conversion-and-loss can be managed. I take the definition for salvage from the Oxford English Dictionary (OED) 'to retrieve or preserve (something) from potential loss'. Of these five ways, there are three that are important ('forestalled', 'recovered' and 'avoided'). Forestalling losses releases those losses that subsequently can go to the same user, a neighbouring user or back to the common pool. Recovery involves the recycling of material losses 'wasted' flowing 'downstream' (physically because of gravity or by time-sequence). Avoided losses are reduced as a part reducing withdrawal and consumption. Avoided losses can flow to same, parallel, upstream, downstream or wider user. The other two means are 'offset' and 'transferred' which move efficiency and production efforts to other systems in order to reduce aggregate depletion of combined systems. The framework uses a conditional 'salvageable' and 'salvaged' when generally correct for the sentence or case. Sharp-eyed pedantics might argue that a distinction is permanently and consistently required because what is salvageable in a future 'intended-to-be-more' efficient system is not necessarily salvaged when that future transpires. These differences add to the complexity of efficiency.
Liminal and liminality	Terms that express 'in-betweenness' or being on the threshold or transiting a threshold. The English Oxford Dictionary definition of 'liminal': 1) relating to a transitional or initial stage of a process; 2) Occupying a position at, or on both sides of, a boundary or threshold. From the Latin *limen* meaning threshold (http://www.oed.com/). In this framework, the idea of 'multiplicative liminality' is employed for the manner in which a wide number of outcomes real or potential are possible when efficiency interventions are applied to natural resources.
The paracommons (as a collective broad term)	Resources with marked efficiency features endow a social-ecological system with unpredictability. Through efficiency changes, resources are on the threshold of resolving themselves into different locations, pathways, fractions, destinations, goods and services. The phrase 'reworking the paracommons' signifies attempts to raise efficiency throughout systems connected by efficiency, loss, and recovery. The paracommons contains five key ideas: complexity surrounding efficiency; three types of liminal paracommons; implications for resource conservation; competition for these gains and interconnectivity between users and systems; and uncertainty surrounding policy and outcomes.

continued …

Table 2 continued

Three liminal paracommons	The framework identifies three types of liminal paracommons, associated with socio-ecological systems undergoing efficiency changes, that can arise during changes to productivity and efficiency, including new destinations for wastes and wastages. The three are: *Productive-consumptive paracommons.* A transitional field of possible efficiency outcomes of material and energy transitions, resulting in unpredictable volumes, amounts and rates of conversions of resources to goods and loss fractions with a paradoxical tendency towards greater aggregate resource withdrawal and consumption. Pathways, locations and forms are not generally specified. The Jevons Paradox is an example. *Multipath paracommons.* A space of unpredictable pathways taken by resources, goods and multiple loss, waste and wastage fractions that leads to different outcomes such as beneficial and non-beneficial consumption or returned to the common pool. Water in irrigation systems is a good example. *Multiservice paracommons.* A space of possible outcomes referred to as ecosystem services. Commonly, several stages of other conversions may be involved. Forests, entailing carbon, carbon dioxide and biodiversity, provide a good example. The framework mostly discusses the multipath paracommons.
Parageoplasia	This describes the phenomenon of one system driving physical changes in other systems remote to the system. This word is coined (Lankford, 2008): 'para' meaning beyond; 'geo' meaning earth or land; and 'plasia' meaning something made or formed. The term is inspired by the concept of 'paraneoplasticity', derived from medical research into cancer, which describes how in the body other cancer-related tumours start to occur remotely from the first and main tumour.
Parasystem	A coined word to describe a 'system of subsystems': a connected system of four types of resource systems (proprietor, neighbour, common pool and wider economy) joined by the efficiency and inefficiency of resource use in the dominating proprietor system. Changes in the loss fraction in the 'proprietor' system impinge on and affect the other users and systems in different ways. The parasystem covers both intra-system users in individual irrigation systems, the hydro-ecology (e.g. wetlands) affected by efficiency changes, the inter-system users across all irrigation systems and water users elsewhere in the river basin. Furthermore, a parasystem describes users connected by the expectation that a policy to raise efficiency will bring benefits to them.
Paragain and delta tare (or salvaged losses)	A paragain is a resource freed up by a combination of a reduction in net demand plus one or more of the five means of salvaging resource losses. A paragain is thus the material gain from changing the efficiency performance of one individual or combined proprietor systems, implying that this 'paragain' can be directed to a user or use. The prefix 'para' signals uncertainty about the salvageability of the gain and its eventual size, location, timing and destination/ownership.
Principal commons	To distinguish the paracommons from the commons (e.g. fish, forests, water, etc.), I have applied the adjective 'principal' (OED, 2012; 'Of a number of things or persons, or one of their number: belonging to the first rank; among the most important; prominent, leading, main.')

Productivity and efficiency	In an aspirational sense, the term 'efficiency' covers efficiency and productivity (in other words that society is aiming to use resources more efficiently and productively). When relevant, specific reference is made to productivity.
Proprietor system	In the REC framework, it is useful to identify one 'proprietor' system (van Halsema and Vincent, 2012) which is the focus of attention and drives the behaviour of the parasystem. A proprietor system is where resources are being withdrawn and converted to products. A proprietor is also scalar (or fractal). Thus a proprietor system might be a single irrigation farm if being discussed in relation to neighbouring farms, or a collection of farms if the latter is being discussed in reference to a neighbouring city. This proprietor system may have its own constituents and clients (for example, a proprietor irrigation system may comprise a collection of farmers).
Resource use efficiency (REC)	A broad term that both captures dimensional productivity measures such as water use efficiency (goods produced divided by volume of water used (consumptively and non-consumptively) and dimensionless efficiency ratios.
Resource efficiency complexity	A theory of complexity applied to resource use efficiency, recovery and recycling. REC has 20 sources which in turn shape and define three types of 'paracommons' where outcomes of attempts to improve efficiency are unpredictable.
Net demand reduction	Net demand or beneficial consumption reduction is achieved by one or more of retrenchment, dematerialisation and substitution.
Retrench and retrenchment	To distinguish between a reduction in withdrawal (and consumption) arising from a) a reduction in losses and b) a reduction in net underlying demand, I use the word 'retrench' for the latter. Unlike dematerialisation, retrenchment is reserved for a reduction in the number of units that give rise to cumulative net demand. An example is a reduction in the number of hectares of irrigation. OED definition (2012): 'to reduce the extent, amount, or number of; to diminish, lessen, cut down.'
Dematerialisation	A reduction in net natural capital consumed per unit of production. Examples include deficit irrigation when for example, 900 mm of water is consumed rather than 1200 mm. OED definition (2012): 'To deprive of material character or qualities.'
Substitution	A reduction in consumption can be achieved by substituting one good or one part of a good for another. For example, to reduce water demand from irrigation at the field level, a lower-transpiring variety of crop might be selected, or a different crop altogether or a different product (e.g. fish-farming instead of irrigated crops).

continued …

Table 2 continued

Loss, waste, wastage, tare, extrinsic fractions	The terms 'loss', 'losses', 'waste/wastage', 'tare' and 'extrinsic fraction' are synonymous. This array of terms conveys the sense that exact scientific terms in the topic of resource use efficiency are difficult to apply to all circumstances because of linguistic usage and understanding. 'Losses' can be divided into wastes and wastages which in turn can be distinguished. Wastes are often seen as undesirable by-products but are usually recoverable and physical and become valuable. Wastages are those losses during conversion that are difficult to capture and recycle (e.g. noise, gas, vibration, heat) – but can be captured with technological innovation. In this framework, I use the terms 'tare' or 'extrinsic fraction' to cover losses/waste/wastage. Thus, 'net' and 'tare' are synonymous with the terms 'intrinsic' and 'extrinsic' respectively.
Gross and aggregate	'Gross' is the addition of an intrinsic fraction to the extrinsic fraction for one given system withdrawing a resource. 'Aggregate' is employed for calculations of combined systems (whole parasystem) useful for when offsetting and transferring efficiency between individual systems. For example, aggregate withdrawal or aggregate classical irrigation efficiency.
Demand, use and usage	The words 'demand', 'use' and 'usage' are general catch-all terms that describe the taking and converting of natural resources into goods, services and consumption. Thus unless clearly stated, 'demand', 'use' and 'usage' mean both withdrawal and consumption. However, efficiency complexity recognises that an inability to distinguish between various types of use leads to improper accounting of resource use. These terms, and their underlying accounting theory and definition require additional work, but the paracommons framework accepts the following: *Withdrawal*: water abstracted from a river or aquifer to meet both consumptive and non-consumptive uses. Some of this water is recycled within the basin system. *Consumptive uses*: water lost from the system principally through evaporation, transpiration and evapotranspiration. Consumptive use is divided into beneficial and non-beneficial for ease of categorisation but in reality these merge seamlessly into each other. *Non-consumptive uses*: water withdrawn from a system but not consumed consumptively – for example, much water use in the household is of this type. This water divides into the recoverable and non-recoverable flows. *Depletive uses*: this is an accounting decision that focusses on the final outcome or disposition of water at the basin scale, choosing not to examine intermediate stages of recycling and recovery. Depletion covers all consumption and other fractions that are put beyond consumption or recovery to the common pool. *Through-flow use*: water not withdrawn but is required in-stream for various environmental and run-of-river purposes such as hydropower generation. For most purposes non-consumptive use and through-flow use can be viewed as synonymous.

The terms above illustrate the complexities of water usage and efficiency – and are under constant review and debate.

Chapter 2

Main introduction and the scope of the book

The study of the commons continues to produce metaphors and labels to better convey understandings of natural resource governance. Updating Hardin's 'Tragedy of the Commons' (1968) examples include: 'commons and anticommons' where under or over regulation is tested (Heller, 1998; Brede and Boschetti, 2009); 'inverse commons' (Raymond, 1999) where greater consumption and sharing leads to greater good (as with open source software); 'new commons' (Hess, 2008) identified as those without developed rules and institutions; 'invisible commons' (Bruns, 2011) covering the challenges of groundwater; and the 'semi-commons' where overlapping ownership regimes in water exist (Smith, 2008). I use the term 'paracommons' to capture emerging uncertainties, complexities and property rights associated with managing efficiency, losses, wastes and wastages applied to resource use conservation.

The paracommons views social-ecological systems being shaped by efficiency and the material salvaged losses from efficiency improvements, giving rise to a wider 'system of systems' than is usually recognised. The paracommons is characterised by resource efficiency complexities associated with the number of potential options and pathways that arise at different scales during efficiency-centred attempts to produce more goods and services from fewer environmental resources. I argue these potential options, moreover our conceptions of these options, and their unexpected decomposition into different outcomes, both guide and thwart resource sustainability and governance. As is explained, this feature of unexpectedness and its bearing on post-efficiency distributions of resources is captured by the concept of liminality or in-betweenness. I conclude that a more comprehensive understanding and response to the challenges of the efficiency and productivity is captured by 'resource efficiency complexity' and the paracommons.

Building on Figures 1, 2 and 3 in the previous chapter, four more figures (5 to 8) introduce this book on resource use efficiency. In a context of growing scarcity, increasing concerns for the distribution of limited resources and rising interests regarding resource use efficiency, interventions are drawn up that aim to increase efficiency and productivity. Yet an arrangement of unpredictable factors and properties captured by the term 'resource efficiency complexity'

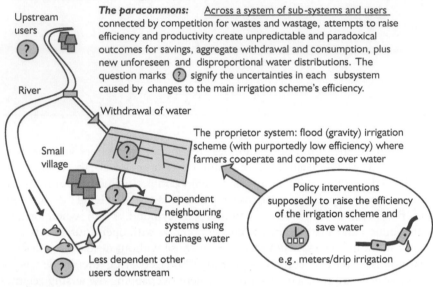

The commons: Within each sub-system, users face rivalry and subtractability over common pool water from: 1) a river; 2) canals within an irrigation scheme; 3) drainage flows recovered from the irrigation system.

The paracommons: Across a system of sub-systems and users connected by competition for wastes and wastage, attempts to raise efficiency and productivity create unpredictable and paradoxical outcomes for savings, aggregate withdrawal and consumption, plus new unforeseen and disproportional water distributions. The question marks (?) signify the uncertainties in each subsystem caused by changes to the main irrigation scheme's efficiency.

Upstream users

River

Withdrawal of water

Small village

The proprietor system: flood (gravity) irrigation scheme (with purportedly low efficiency) where farmers cooperate and compete over water

Policy interventions supposedly to raise the efficiency of the irrigation scheme and save water

Dependent neighbouring systems using drainage water

Less dependent other users downstream

e.g. meters/drip irrigation

Figure 5 Irrigation systems in river basins: the commons and paracommons

(REC) strongly associated with recycling and recovery introduce a great degree of uncertainty to this policy process. In turn, efficiency connectivity, complexity and policy uncertainty give rise to the paracommons.

In Figure 5, attempts to raise the efficiency of a large irrigation system not only leads to paradoxical and complicated outcomes for the irrigation system, but also to paradoxical and unforeseen outcomes for whole river basin and upstream and downstream systems. While 'the commons' frames these component systems as each having their own common pool problems, the paracommons frames these systems juxtaposed in parallel connected by efficiency and recycling. The paracommons tells us that 'savings' of water by, for example, a change from flood/gravity irrigation to drip irrigation (yet both not accurately assessed for their 'efficiency' and therefore subject to opinions on losses and consumption) cascade in different ways to different farmers within the irrigation system, farmers recovering drainage flows, dependent freshwater ecologies and downstream and upstream users.

Figure 6 introduces the seven main issues that make up the paracommons and its context. The top 'box' of Figure 6 gives the first – capturing a number of drivers of new approaches towards efficiency. A changing context of increasing scarcity, resource recycling, allocation/ reallocation and green growth establishes new urgencies that drive up an interest in the role of efficiency in addressing

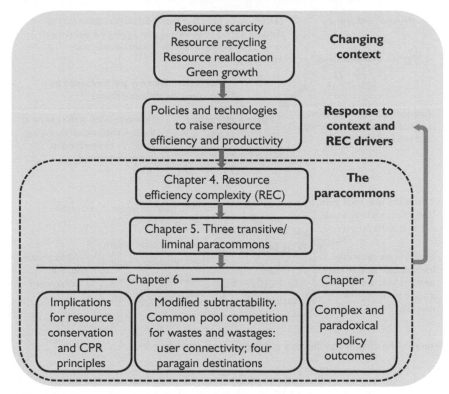

Figure 6 Resource efficiency complexity and the paracommons

these (Keys *et al.*, 2012). In the second box, policies to improve efficiency and productivity are constructed in response to these changing circumstances. Policies need not be only formal programmes designed by funders, but also can be revealed through new agreements by farmers sharing a small tertiary canal. These policy intentions reveal desires to improve efficiency to a higher standard of performance at some point in the future.

At the bottom of Figure 6, a combined group of five issues makes up the core concepts of the paracommons. These are: 1) the complexities of resource use efficiency (REC); 2) systems in transition giving rise to liminal 'spaces'; 3) implications for resource conservation, regulation and CPR principles; 4) competition over the losses/waste/wastage between users and systems and users interconnected via the relocation of salvaged losses, and; 5) disparities between policy intention and outcomes resulting from a lack of appropriate understanding, measurement and monitoring of resources.

Building on Figures 1 to 3 in Chapter 1, Figure 7 introduces the derivation of an 'efficiency material gain' generated by the expectation of a future efficiency performance being higher than an assumed lower current efficiency. In Figure 7,

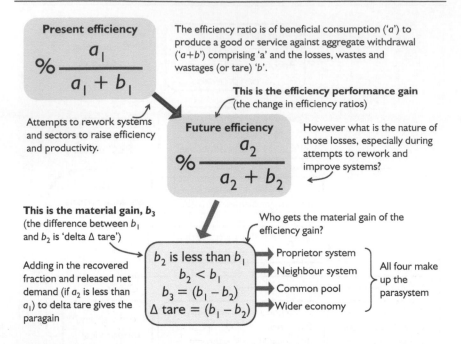

Figure 7 The idea of an efficiency gain leading to a material gain

a higher efficiency in the future results from an intended tare reduction (losses b_2 are lower than losses b_1).[1] If it can be captured and recovered, this 'efficiency gain' comprising 'delta (Δ) tare', equal to $(b_1 - b_2)$, can be directed to different destinations, uses and users. However, not shown in Figure 7, but discussed later, is that losses in the future system are potentially already available (although changing) by being recovered. Furthermore, the net demand might be reduced (for example, in deficit irrigation). Thus the reduced demand, reduced losses and the recovered losses must be added together – creating a fuller picture of salvaged resources (termed here the 'paragain'). By asking who then gets the paragain from an efficiency/productivity gain, the idea of competition over salvaged resources arises. In Figure 7, and elsewhere in the chapter, four main destinations or parties are identified; the proprietor making the efficiency shift, the proprietor's closely connected neighbours, the common pool that supplied the resource and other users (or other systems) in the wider economy. (Note: detailed explanations of gain calculations are found later.)

Figure 8 shows these two types of 'losses' (recoverable/recovered and forestallable/forestalled). The left-hand photo shows water physically spilling from a final check-gate in an irrigation canal. This 'wasted' water passes down the canal, now acting as a drain, back into a river some few kilometres away.[2] This water can be recovered and reused by downstream uses and users. This is different from what is happening on the right-hand side of Figure 8. Here,

(a)   (b)

Figure 8 Irrigation losses that can be recovered or forestalled

Photos show (a) leaking stop-gate in an irrigation canal leading to drainage water recoverable downstream in a nearby river (but not easily recoverable by next field in the sequence of irrigation), and (b) and standing water in a recently harvested rice field leading to non-beneficial evaporation not recoverable at this moment (but can be reduced in future).

the open water in the field after a rice crop has been harvested evaporates non-beneficially. It cannot be recovered in 'the here and now' (because the water sits in the harvested field without draining) but it can be reduced/forestalled. To forestall this 'wastage' (non-beneficial, hard to recover consumption) requires the rice farmer to take action either in the same field next year or in neighbouring fields still to be harvested. To reduce/forestall these losses, the farmer would close the canal gate leading to the field some 10 or 20 days prior to harvest so that the rice crop transpires the standing water and soil water depleting it in time for harvesting. Many crops ripen healthily using remaining soil water without needing an irrigation in their final growth stage. This would mean future harvested fields not having standing water in them that then evaporates. This reduced wastage can now be kept back and released for use elsewhere – either in the irrigation system or in the river system.

While Figures 1 to 8 fulfil an introductory function for this chapter, as will be seen, resource use efficiency is complex, giving rise to considerable room for misunderstanding and error in terms of measurement and policy. Furthermore, efficiency as a performance indicator creates the implicit expectation that it is to be raised by attempting to re-work and improve the technologies and management of systems – though the means and consequences of that may not be understood. Murray-Rust and Snellen (1993), in discussing irrigation performance, explain on page 7: 'Performance indicators, by providing information on past activities and their results, help in making informed judgements which may guide our decision making about future activities'. It is possible to interpret this view more critically. Not only do performance indicators help guide decision-making, they appear to encourage the judging of systems, often without measurement (or without accurate measurement), in turn creating the conditions for a perceived efficiency deficit between the system's harshly hypothesised current performance and the future system's expected step-up in performance.[3]

Thus the paracommons can be seen as a 'politics of expectancy' (c.f. the political economy of promise; Leach *et al.*, 2012) framing the uncertain differences between; a) the prefigurations of the promise of efficiency and productivity gains and; b) the extant and often unforeseen material, productive and distributive physical and social outcomes for users and resources following attempts to make savings.[4] These differences arise because efficiency-type conversions at different levels (e.g. field, scheme, basin) are multiplicative. They lead to numerous possible outcomes when boundaries, regulation, accounting and measurement are poorly controlled or deployed, and because too often the science of efficiency improvements – either theoretically or in practice via technological or institutional interventions – is inadequate. While some of these ideas are captured in the literatures around the Jevons Paradox (Polemini *et al.*, 2008) and agricultural water savings (Seckler, 1996; Crase and O'Keefe, 2009), this book provides further analysis of this phenomenon. Although carbon, forests and energy are discussed, I mostly select irrigation to discuss the paracommons because: the wastage fraction is volumetrically large and valuable; irrigation water 'flows' via multiple pathways to different outcomes which are difficult to measure; savings in irrigation are subject to competing legal claims; irrigation systems increasingly dominate water use patterns in arid and semi-arid river basins, and; the sector witnesses uncertain applications of technologies to drive savings.

2.1 A rising interest in scarcity – and efficiency?

In a 7-billion plus world exploring the limits of resource scarcity, availability and distribution, the science and politics of resource efficiency and performance dictate the extent to which living and environmental standards may be maintained, made equitable and just. Sax (1990, p.258) wrote in reference to resource limits on 'spaceship earth':

> It is not by accident that we are turning towards the control of waste and water marketing as ways to reallocate existing supplies and meet new demand. There is also increasing interest in reuse of existing water supplies and in technical means to achieve equal output with smaller inputs of water.

More recently UNEP (2012, p.17) wrote 'More efficient use of all resources in all sectors can help reduce water demand and water pollution, thus alleviating pressure on the environment and related ecosystem services'. These challenges have expressed themselves in a variety of efficiency and ecological thinking that has arisen in the last 20 years; eco-efficiency, industrial ecology, industrial metabolism and x-factor production (Socolow *et al.,* 1996; Reijnders, 1998; WBCSD, 2000; Anderberg, 1998; CIAT, 2012); ecological modernisation theory (Warner, 2010; York and Rosa, 2003) and green growth (OECD, 2011).

However, despite more attention being brought to bear on resource performance driven by fears about scarcity, the position of efficiency in this debate appears to be a decidedly mixed affair. As explored in Chapter 4, various literatures miss or misinterpret efficiency. It is my view that the physical and material dimensions of efficiency do not have a well-understood place in attempts to resolve imbalances between supply, demand and sharing of resources. Its present ambivalent status is curious – but understandable if scientists intuit that physical efficiency is not necessarily 'the solution' given the multiple variables that drive the consumption and distribution of resources or that an economic treatment of efficiency (which is receiving more attention – see Bleischwitz *et al.*, 2011) can drive technical and material innovation without examining how that might transpire at both wider and local scales.

Ideally, the pressures to give efficiency a higher status are in place. Increasing demand, scarcity, finance accountability and interconnections between input resources (such as land, energy, water and labour) establish new impetuses for management. These are: to reduce consumption of resources while maintaining economic growth; to save money, labour and other inputs; to reuse and profit from waste products; and to reduce harmful pollution. Three interrelated responses should arise – the first is that wastage and waste as a process (e.g. irrigation efficiency) and as a material product (e.g. drainage effluent) come under increasing scrutiny. Second, attempts to manage the natural resource 'commons' by reducing and recycling waste/wastage become more attractive (Jones *et al.*, 2011) and third, measuring waste and wastage becomes critical to knowing and adjusting the management of efficiency.

But these three responses are not straightforward. Complexity defines the subject area of efficiency. A range of conceptual and practical questions arises regarding how we perceive the role of efficiency in resource management. Examples include the extent to which we discern processes of recycling of waste and the selection of boundaries of systems under scrutiny. Scaling up to larger systems in efficiency science is a challenge; it is one thing to decrease the water consumption of a farmer's field, but it is a different matter to reduce the impact of an irrigation system on its hydrological environs or at a larger scale, to improve the efficiency of a country's irrigation sector (as Spain attempted in the last 12 years with mixed results, see Lopez-Gunn *et al.*, 2012). Further complications in efficiency science arise from and apply to the claims (Larson and Richter, 2009) that water losses can be 'saved' in one location in order to offset consumption elsewhere which requires an even higher burden of evidence.

The quantitative dimensions of resource inefficiency introduce another reason to be interested in efficiency. In some sectors, losses might be significant enough to be a solution to scarcity. One way to show this is to compute the amount of resource 'lost' in irrigation (an expression I employ for illustrative reasons). Globally irrigation systems are widely cited to be 40 per cent efficient (see Lankford, 2012b, for a further discussion on this commonplace narrative) thereby 'wasting' 60 per cent of freshwater, in other words, the potential gains to be had for water allocation from even meagre 'savings' of water, are supposedly large. A calculation hints at the potential material gains from raising irrigation efficiency –

subject to the provisos in this book. By assuming a global irrigated area of approximately 280–300 million hectares, of which approximately 85–90 per cent is gravity/surface fed, we could for the purposes of demonstration, accept a 10 per cent relative reduction in total consumption (via non-beneficial consumption and non-recovered losses). Assuming a cautiously low gross annual consumption of 600 mm (building on Döll and Siebert's (2002) figure of approximately 420 mm net crop water requirement globally) this 10 per cent saving in consumption gives a reduction of consumption down to 540 mm, releasing 60 mm depth equivalent. Spread over 280 million hectares, this is equivalent to 0.46 cubic kilometres water per day, the same volume as providing 7 billion people with approximately 65 litres per day of water per person, a sizeable proportion of an individual's daily water requirement of say 60 to 200 litres per day depending on circumstances.

Although this calculation reveals the considerable volumes potentially involved, this framing of 'savings' hints also at the contentious nature of the debate. While careful not to ascribe a low absolute efficiency to surface irrigation of 40 per cent (commonly done), instead assuming a reduction of 10 per cent in consumption (again an assumption), the calculation presumes that water can be 'saved' then extracted from one sector and then geographically and hydrologically moved to other sectors without significant transaction losses or appropriation. This concatenation of logic, ignoring contested, appropriative environments and the absence of sound accounting, should be treated critically. Nevertheless, the volumetric sketch reveals the political attraction of raising irrigation efficiency – the sector seemingly 'misuses' large amounts of scarce freshwater that could be better utilised, it is believed, elsewhere.

Finally, the idea of the paracommons stems from the observation that 'today's' efficiencies, losses, wastes and wastages are managed and governed poorly. In other words, society holds views about what should be done about current 'inefficient' systems and goes about managing those systems using a variety of interventions, technologies and institutions. That society misunderstands the nature of efficiency and how to manage losses, and argues that higher efficiency will necessarily reduce environmental impact (or the opposite that it paradoxically will always give us greater consumption) is an integral part of the paracommons.

Furthermore, the paracommons goes beyond resource 'accounting' (Godfrey and Chalmers, 2012; UNSD, 2007). In other words as a substantive point, the paracommons views water accounting as one part of managing water systems' performance (Foster and Perry, 2010). Accounting at the basin level, does not a) explain why and how water flows to different outcomes; b) it cannot guide the myriad local switches, bifurcations and actions that result in cumulative outcomes; c) it can only partially comment on efficiency policy; and d) it cannot interact with people's views on local efficiency. In the parlance of financial accounting, I contrast end-of-year 'statutory' accounts interested in final outcomes (which is what most water accounts seem to focus on) to two other types of accounts: regular month-by-month 'management accounts' which (in different forms) interest irrigation managers and farmers in order to run their systems successfully and efficiently and 'appropriation accounts' which determine who gets the year-end profits.

2.2 Complexity and common pool resources

Research on environmental issues increasingly recognises that complexity is a defining character of the sustainability, conservation and governance of common pool resources (Scoones, 1999; Manson, 2001, 2008; Underdal, 2010) and that, if anything, this complexity appears to be increasing not only through more refined understandings of nature-human interactions (Norgaard, 2010) but because scarcity, innovation and rising populations disturb the balance of environmental protection and economic development (Tainter, 2011).

In this book I argue that engaging with sustainability via efficiency imbues social-ecological systems ('SES'; see Cumming, 2011; Folke, 2006) with complexity and unpredictability. Therefore the REC and paracommons add to other risks and complexities in resource science and governance where outcomes either do not meet expectations, are unpredictable or difficult to manage. Other risks include random variation, measurement error, and non-linearity in the natural world (Clarke, 2006): dynamic equilibriums (Leach *et al.*, 2010); non-stationarity and natural variability (Lundqvist, 2009); inappropriate frames, models and indicators of understanding (Jerneck and Olsson, 2011), and inappropriate governance regimes ill-fitted to ecosystem types (Berge and Van Laerhoven, 2011).[5]

In addition, a distinction is made between common pool governance efficiency (Section 3.2.3) and resource use efficiency. The former sees complexity in the efficiency of economic and institutional arrangements that manoeuvre common pool resources towards sustainability (Singleton, 1999; Bretschger, 2011). This is a concern with designing appropriate institutions including pricing mechanisms that work effectively on the problem of common pool governance. On the other hand, the topic of this book sees complexity in the technical and material dimensions of resource use efficiency. This book is concerned with the changes in physical transformations of common pool abstraction into goods and services. Regrettably, for lack of space (and field data) the two types of efficiency are not brought together.

At this point I can also identify five water-specific complexities other than efficiency which shape flows and boundaries and establish problems of ownership, excludability and rivalry. In the first, water is subject to multiple uses and users; navigation, hydropower and irrigation being three examples. Second, some uses are consumptive (irrigation) while others are non-consumptive (navigation). Third, water is fugitive and moves through the landscape, so subtraction in one location can diminish a resource distantly. Subtraction can occur via consumption or via adjustments to switches between bifurcating networks. Fourth, water exhibits unpredictability and variability in both demand and supply. Finally water availability is further shaped via a number of properties, such as timing, location, quality and quantity. All five establish vexed questions over the derivation of workable property regimes related to monitoring and sanctioning demand against supply. By way of example, policies to privatise water, as witnessed, for example, by the sale of water rights to irrigators, influences and is influenced by these complexity properties.

2.3 Efficiency, savings, gains, paragains and uncertainty

The difficulty in precisely discussing resource efficiency is that a reduction in natural capital consumption via reducing losses cannot be isolated from a reduction in the production of goods that in turn reduces consumption and losses. In other words, one can reduce gross demand by reducing both the losses (tare) and underlying net demand that gives rise to losses. Thus although the words 'net', 'gross' and 'tare' as nouns allow for some measure of precision (to reduce gross demand, one can reduce either the tare or the net demand), they do not convert easily into verbs and actions. Furthermore in 'real' systems comprising canals, fields, soils and drains, they overlap, making attribution difficult. Lack of definition is reason enough for terminology to be selected as one of the 20 resource efficiency complexity factors discussed in Chapter 4. However, given this confusion, I include here a simple framework of ideas that underpin the balancing of supply and demand (Table 3).

First, the top part of Table 3 argues that a deficiency in supply can be addressed by 'supply management' (see Section 3.2.2). Second, the lower part of Table 3 gives four 'demand management' pathways to reduce demand for natural capital; by 'salvaging' losses; by dematerialising the goods produced; by retrenching the number of goods or component parts produced, and by substituting goods or component parts for other goods or parts that consume less of a resource.

The main scope of the book addresses efficiency through the salvaging of the loss through five different means (discussed in Chapter 4). However, in irrigation the subject of deficit irrigation (a form of dematerialisation), not planting up and irrigating areas (retrenchment), and selecting a crop type or variety that consumes less water (substitution) are so closely allied to concerns about water productivity and 'saving water' that these topics are also addressed. Thus in Table 3 the freed up paragains come from a mix of salvaging losses (tare reduction), dematerialisation, retrenchment and substitution (net demand reduction) when appropriate.

However, a separate substantive policy discussion about dematerialising, retrenching and substituting net demand is not the topic of this book. Moreover, these three ideas might be interpreted as an acknowledgement that efficiency is either of minimal consequence or impossible to effect (in other words; don't address efficiency but simply reduce irrigation command areas). This is not the stance of this book; instead the reduction of net demand is wrapped up closely with the reduction of losses. Nonetheless they imply intractable cultural, political and economic decisions that affect the original sources and amount of

Table 3 Addressing sufficiency (supply is less than demand)

Main idea	Part	Sub-type	Means	Example or explanation
Supply management to boost or switch supply			Various	Building a dam or desalinising seawater
Demand management (reduction in gross demand) Gross = tare + net	Tare, loss	Salvage	Forestall Recover Avoid Offset Transfer	Reduction/release of losses Recovery of losses Avoidance of losses by reducing net demand Offsetting of losses Transfer of losses
	Net demand (or beneficial consumption, BC)	Dematerialise	Volume Density	Reduction in size, density, weight of units, goods and services that give rise to net demand (or amounts of material consumed in per unit production). An example is deficit irrigation. Smaller loaves of bread exemplify weight dematerialisation and less nutritious or less dense bread represent density dematerialisation.
		Retrench and retrenchment	Number	Reduction in number of units, goods or services that give rise to cumulative net demand. Example is number of hectares of irrigation or the number of bread loaves purchased and eaten.
		Substitution		A number of approaches exist depending on which part of the chain of production is substituted. Effects can be very similar to dematerialisation and retrenchment.

demand – and how to value and account for resources not taken (Gavaris, 1996). Examples include decisions over the total number of houses built in a given area (related to plot size and density which constrains garden size and watering), or publicity efforts to reduce garden watering during a drought or to change a lawn to a gravel bed.

Figure 7 can be seen at work in an example given in Table 4 which shows a single irrigation system of 1300 hectares responding to an efficiency programme plus two types of net demand reduction. The efficiency improvement programme takes its dimensionless efficiency from 55 per cent in column A to 65 per cent in column B. In column A, if its net water requirements arise from a crop water requirement of 750 mm, then 55 per cent efficiency converts a net volume of 9750 m³ required for the 1300 hectares of into a gross volume of water of 17,727 m³ – the difference being the 'tare' fraction, the losses of 7977 m³. Crucially it is the gain in efficiency performance from current status (column A) to a future status (column B) that 'frees up' 2727 thousand cubic metres, or the difference between the greater tare (loss) in the current scenario and the smaller tare fraction in the future. This 'delta tare' makes up part of the 'paragain' referred to in this book and is the resource potentially newly available for competitors to compete over or share.

In column C the effect of a reduction in net demand via dematerialisation (deficit irrigation) on reducing withdrawal (gross demand) is revealed. The net crop water requirement of 750 mm drops to 600 mm which, retaining an efficiency of 55 per cent, leads to a reduced gross demand of 14,182 m³. This frees up a delta tare of 1595 m³. However, the total gross demand has dropped from 17,727 m³ in A to 14,182 m³ in C which releases 3545 m³, which is greater than the released 1595 m³ only associated with the reduction in losses.

In column D, the effect of retrenchment (areal reduction) is calculated. A drop in area from 1300 to 1000 hectares reduces the net demand from 9750 m³ in the original scenario down to 7500 m³. This salvages 1841 m³ from a reduction in losses (delta tare), but actually 'frees up' 4091 m³ because net demand has decreased by 2250 m³ (1841 plus 2250 gives 4091).

However, a more accurate computation should also allow for recovered flows wrapped up within the 'tare'. For this reason, greater detail on paragain computations are given in Tables 13 and 22 where different means of salvaging losses are included. In addition, these 'gain' computations are framed by how scarce and variable the resource supply is. For example, water on the supply-side is scarcer than, say, atmospheric carbon dioxide which effectively is not only unlimited but is being added to. Furthermore, scarce water utilised in or released from irrigation systems physically affects water volumes for users many miles away – yet abundant CO_2 captured in or released from one forest does not subtract from another forest. Thus in the case of water being converted to irrigation evapotranspiration, the efficiency gain creates a material gain (the paragain) that is mobile and competed over. In the case of carbon dioxide converted to carbon, the efficiency gain that delivers more carbon, while important does not lead to stringent and distant competition; there is still much carbon dioxide left to convert.

Table 4 Performance gain giving a material gain (and examples of net demand reduction)

	Computation of row numbers (for columns A and B only)	Current efficiency of system	Future efficiency of (salvaged loss)	Future demateriali-sation	Future retrench-ment
		A	B	C	D
			Tare reduction	Net demand and tare reduction	
Area of irrigation system (ha)	1	1,300	1,300	1,300	1,000
Crop water req. (mm depth)	2	750	750	600	750
Net water demand (000 m³)	3 1×2/100	9,750	9,750	7,800	7,500
Efficiency	4	55%	65%	55%	55%
Gross water demand (000 m³)	5 3/4	17,727	15,000	14,182	13,636
Tare/loss (000 m³)	6 5–3	7,977	5,250	6,382	6,136
Delta tare (000 m³)	7 A6–B6	–	2,727	1,595	1,841
Delta net vol (000 m³)	8 A3–B3	–	0	1,950	2,250
Delta gross vol (000 m³)	9 A5–B5 or 7+8	2,727		3,545	4,091

Note: Gross demand = (net crop water requirement * area) / efficiency

2.4 On the word 'paracommons'

I have coined the word 'paracommons' to capture ideas of complexity and liminality surrounding resource use efficiency. The previous chapter gave the etymology of 'para' drawing from ancient Greek meaning 'by the side of', 'beside', 'hence' 'alongside of', 'by', 'past', 'beyond'. Also 'para' expresses 'to one side', 'aside', 'amiss', 'faulty', 'irregular', 'disordered', 'improper' or 'wrong'. Furthermore, 'para' portrays subsidiary relation, alteration or comparison and covers 'miscellaneous terms in the sense analogous or parallel to, but separate from or going beyond, what is denoted by the root word' (OED, 2012).

There are six related ideas in the paracommons signalled by the prefix 'para'. Table 5 gives more information on these interpretations. First, the idea of 'para', meaning 'altered', 'against' or 'contrariness', is employed and is contained in the word *paradox*. The paradoxical outcomes of the paracommons (of which the Jevons Paradox is an example) arise from the complex, unknown and poorly controlled resource pathways decomposing to different outcomes, in turn suggested by the core concept of paracommon liminality (a transitive threshold containing different pathways and outcomes). The prefix 'para' implies counter-intuitive or counter-to-expectation outcomes – for example, that an increase in efficiency leads to greater depletion of natural capital. 'Paradox' ('against doxa') reminds us that prior to raising efficiency we (science and society) hold expectations and beliefs about intended outcomes (see, for example, 'the political economy of promise' Leach *et al.*, 2012).

Next, three spatial ideas are invoked when 'para' denotes 'alongside' or 'beside': a) closely connected neighbouring users or systems *within the same level* (as in *parallel*); b) closely connected neighbouring users or systems from a *subordinate dependence* point of view (as in *parasymbiosis*), and; c) less closely connected users *above a level* via 'remoteness' (as with *paraneoplasm*). Table 5 explains these three levels (P2, P3, P4) in more detail as does further discussion ahead. The paraneoplastic term comes from medicine, describing the occurrence of cancer growths elsewhere in the patient remote from the original cancerous infection. In Lankford *et al.* (2009), I coined the word 'parageoplastic' as its geographical equivalent, describing remote and loosely connected externalities driven by upstream water depletion (para = beyond, geo = place and plasia = forming). Furthermore, the coined word 'parasystem' conveys a wider system of systems arising from neighbourliness and nestedness.

The fifth idea is when 'para' means an 'abstract alongside' (as in *paraphysical*). Thus, the paracommons sits in an abstract form alongside the commons.[6] As explained in the book, the paracommons continuously physically decomposes or resolves into the principal commons when flows of resource-and-losses pass through one of three types of paracommons (described in Section 5.3). In other words, the paracommons expresses the pre-efficiency and post-efficiency changes to the whole parasystem but users in different parts of the parasystem continue to physically 'see' their situation normally as 'the commons'.

Table 5 Six concepts denoted by the prefix 'para'

Key concept	Dictionary definition and example word	Application to the paracommons
Contrariness Altered	(P1) Paradox: an absurd or self-contradictory statement or proposition, or a strongly counter-intuitive one, which investigation, analysis, or explanation may nevertheless prove to be well-founded or true.	Outcomes are counter to expectations and promise. Complexity, uncertainty and liminality of efforts to raise efficiency lead to unexpected performances, high consumption and other distributive consequences.
Alongside – same level. Or 'joining together'	(P2) Parallel: applied to things running side by side in this way, or pointing in the same direction. Parabiosis is defined as the anatomical joining of two beings or individuals.	Raising the efficiency of systems leads to unforeseen consequences for strongly connected users in parallel to each other within a particular scalar level.
Alongside and immediate: subordinate levels	(P3) Parasymbiosis: For example, a relationship in which lichen supports a fungal species growing in close association with it, without apparent disadvantage. A commensal association short of full mutualistic symbiosis.	The paracommons features recycling and recovery involving downstream users and systems strongly and immediately dependent on 'wastes' from upstream systems.
Alongside, beyond, remote, or superior levels	(P4) Paraneoplastic: designating or relating to any of various (often metabolic, endocrine, or neurological) conditions which occur in patients with neoplastic disease but which do not result directly from tissue invasion by the primary tumour or its metastases.	The paracommons features remote consequences for aggregate consumption of water from the whole river catchment, and distributional outcomes both upstream and downstream of the system as a result of efficiency changes ('parageoplasia').
An abstraction, a construct	(P5) Paraphysical: not part of the physical world as it is currently understood.	The paracommons is an abstraction of the commons. The parasystem is an abstract two-axis 'system of systems'.
Relative and changing viewpoints	(P6) Parallax: difference or change in the apparent position or direction of an object as seen from two different points.	Perspectives on the changes to the efficiency of a system come from those dependent on efficiency changes and recovering wastes from that system.

Sixth, the word *'parallax'* reminds us that the perspectives of resource efficiency depend on perspective from where one 'sits', in other words, from the system making the efficiency change; from the viewpoint of any neighbouring dependents (previously satisfied with 'wastes' but now worrying about a reduction in waste flow because of upstream efficiency efforts); and from the wider natural or economic system that might witness greater aggregate withdrawal and consumption. As discussed later, the paracommons invokes questions of boundaries given that the consequences of efficiency changes can be felt at different scales and levels.

To be forthrightly clear, 'the paracommons' is not about waste and wastage *per se*. It does not apply to systems of reuse and recycling when the flow of the resource and the resource system are static and not undergoing efficiency changes. For example, an irrigation system downstream of a city's wastewater outlet does not constitute the paracommons if the flow of wastewater is predictable. This is a standard example of the commons; wastewater is the common pool resource for irrigators in the downstream system. What would turn this situation into the paracommons is when water users within the city begin to implement their own efficiency changes and harvesting of surface water impinging on the wastewater irrigators. This would lead to questions over the ownership and nature of the savings, uncertain redistributions of water resources and new yet-to-be-defined connections between the city and downstream water users.

2.5 On the word 'paragain'

Paragains describe the potential resources that come from changes to the efficiency and net demand of resource use. 'Losses, wastes and wastages' salvaged as a result of improving efficiency help make up paragains alongside any resources that are 'freed up' by substitution, dematerialisation and retrenchment. A paragain is thus a material gain (e.g. water, carbon) rather than 'released cash or time' though the latter may be wrapped up with intentions of an efficiency project. A 'paragain' emphasises the five ideas: the first four of them connected to the idea of a 'new resource'– a resource 'freed up' for reuse. The fifth idea introduces the sense of potentiality – that paragains are in limbo dependent on current and future resource use. These ideas are explained here:

First, a paragain includes recycled resources. The term 'new water' is used in the Middle East (e.g. Mediterranean Water Scarcity and Drought Working Group, 2007) with reference to recycled waste water – capturing the sense that wastes and wastages are valuable. Wastes and wastages generated by a proprietor system can be recycled by other users, and the aggregate system can become less consumptive by increasing the amount of waste/wastage recovered – boosting the amount of 'new' but previously discarded resources. Thus in an irrigation system, a drainage flow that previously went ignored to a sink is now channelled to a small village to provide useful benefits. However, a paragain derived from recovered resources may also be impacted upon by efficiency changes within the proprietor system that reduce the amount of recoverable water. For example, an irrigation system that uses water in its fields more effectively reduces the amount of water spilling into drains giving a lower amount of recovered water. Thus a paragain must be very accurately accounted for and defined.

Second, following on from the previous point, a paragain covers forestalled losses. This is the sense of the 'new water' implied by Gleick *et al.* (2011) in their discussion on where extra water could be found for California. In this case, a forestalling or avoidance of the amount of waste/wastage created within an irrigation system between a currently inefficient system and a more efficient future system forms the released 'new resource' for further use, competition and rivalry. In other words, a paragain is a 'new resource' found by reducing-and-releasing resources rather than recycling losses.

Third, a paragain comes from a reduction in the net demand over two time periods: current demand and a lower future demand – the latter initiated through substitution, dematerialisation and retrenchment (Section 2.3). Thus paragains do not only come from changes in losses, they arise because notions of unwarranted demand are addressed by being careful and parsimonious in other ways – for example, via deficit irrigation. However, losses can be reduced at the same time because all the fractions are coupled together (Section 4.6.5).

Fourth, from the previous three points, paragains reveal the uncertain nature of the paracommons. Substantively, paragains raise the prospect of a commons around the resources freed up when verifiably reducing aggregate demand or recycling of wastes and wastages – but this occurs in complex ways. This opens up the question of 'Who gets the gain from an efficiency gain?' (and its corollary; 'Who wins from an efficiency decline instead of an efficiency increase?').[7] Furthermore, it is not clear at which point in the sequence of events from resources sitting in the common pool through to final disposition that paragains emerge to be competed over. Computations later in book reveal this (e.g. Table 22).

However, fifth, the term paragain moves beyond 'new water' (a term I do not favour as it poorly captures the idea of recovery/recycling of losses, plus literature on mining resources uses the term 'new' for newly mined minerals reserving 'recycled resources' as previously mined and used resources). While a paragain thinks of as a 'new resource' in the sense of Gleick *et al.* (2011), it is a word that signals much more complexity and caution – hence the prefix 'para' is attached to gain. What is currently a loss is not by any means a verified 'new resource' for the future because the act of forestalling losses can also reduce currently recovered losses.

Thus efficiency-induced paragains are in limbo. They are not confirmed as 'new resources' (Gleick *et al.*, 2011) nor refuted as not being accountable as 'new resources' (Perry in Frederiksen *et al.*, 2012). Today's salvageable/recoverable resource is not the same as tomorrow's salvaged/recovered resource. The term paragain also cautions that water as a recycled resource might be indistinguishable as water gained via a forestalled loss (but their management is remarkably different). Referring to this last point, 'paragain' covers all five ways losses may be salvaged (explored in Section 4.5.2).

2.6 On 'liminal' and 'liminality'

Central to the idea of the paracommons and REC are attempts to raise efficiency between a current (and assumed) inefficient system and a future (and expected) more efficient system. These attempts, also aiming to free up a resource, create systems in transition; they put systems on a threshold of change between two states – 'now' and 'future'. I have applied the terms 'liminal' and 'liminality' to the paracommons for the manner in which the terms capture the uncertainty and 'in-betweenness' created by multiple options arising when attempting efficiency/productivity changes. Liminality also applies to the idea that losses sit between a state of being salvageable, potentially claimed by many parties, but then salvaged by a known and specified party – perhaps paradoxically depending on prior expectations.

The term liminality arose through the social studies of van Gennep (1909) who explored rites of passage in various societies. The term has, amongst other applications, described the transitory period between stages of human experience (Buckingham *et al.*, 2006), to change within communities (Lawrence, 1997) and to geographical histories of rapidly changing nation states 'being between positions' (Yanlk, 2011). In these literatures, it is the potential transition-in-waiting, rather than tangible outcomes and new states, that interest scholars.

From discussions and comments on drafts of this book I might conclude that liminality is not an everyday concept amongst natural resource scientists. However, resource use efficiency is socially and physically complex arising from a range of fault-lines and overlaps between and within different disciplines, sectors, systems, and understandings. The idea that liminality applies to social systems but not to technological social-ecological systems is arguably incongruous. In addition, I am intrigued by the idea of the vinculum (a mathematical threshold) at the centre of an efficiency ratio and by the multiple potentials and uncertainties that sit between policy intentions to raise efficiency and 'real' aggregate outcomes. Although *paradox* captures the uncertainties of liminality, I wish to explore more fully beyond the 'paradox' characterisation found in efficiency literatures to consider the distributive elements of efficiency via interrelated systems connected by resource recycling and savings. It is these opportunities for complexity, 'in-betweenness' and 'transition' that suggest liminality. A full exposition of the various reasons for utilising the concept of liminality is given in Section 5.1.

2.7 Purpose, framework and structure of the book

The purpose of this book is to explore how resource distribution, complexity and access in technological social-ecological systems are influenced by efficiency-driven changes to the management of those resources. By reframing (Runhaar *et al.*, 2010) *efficiency as resource* (and, under scarcity, resources mediated by efficiency), the book addresses five emerging, closely connected phenomena (seen in the lower half of Figure 6). Each is located as follows:

- The first is that resource efficiency and efficiency-induced resource re-distributions are shaped by uncertainty and complexity. Resource use complexity (REC) is the term utilised to capture an array of properties of resource systems and efficiency reworking that in turn explain the existence and behaviour of the paracommons. See Chapter 4.
- Connected to the concept of complexity is the idea of multiplicative liminality (or transitive 'threshold-ness') invoking the idea of many possible unforeseen or uncontrolled outcomes not easily predictable prior to making efficiency improvements. In turn, this idea of complexity and liminality suggests that it is not easy to answer the question of who gets or owns the gains made through efficiency improvements. The word paradox captures this phenomenon. Chapter 5 addresses the issue of liminality in depth.
- The third – resource conservation – utilises efficiency to revisit ideas held within the common pool resource (CPR) literature on how institutions self-manage and self-govern the commons in order to militate against over-abstraction. This topic is addressed in Section 6.1.
- The fourth idea examines the 'commons' notion of subtractability and rivalry which no longer applies to resources with inefficient fractions because salvaged losses become available to others. This 'second chance' means that individual resource systems become interconnected by changes in efficiencies and losses – and this connectivity generates a 'parasystem' out of the individual systems. The topics of modified subtractability and interconnectivity crop up in some of the 20 REC factors (Chapter 4) and are returned to in Section 6.2.
- The fifth idea is that complexity and liminality interject into the policy process confounding expectations. This topic is addressed towards the end of the book in Chapter 7.

The paracommons is the identity given to these five ideas, allowing the concept to be contrasted with 'the commons'. In addition to the introductory ideas given in Figures 1 to 8, Figure 9 may be seen as the conceptual framework of resource efficiency complexity and the paracommons. The outer ring provides the shifts in context that set the scene for resource use efficiency. Issues such as heightened resource scarcity reside here. Located within the ring are 20 sources of resource efficiency complexity (left-hand side of Figure 9) which give rise to three paracommons (the centre of Figure 9) as key defining features of the paracommons. In addition, resource efficiency complexity shapes the other five phenomena listed above and in particular inserts itself between policy intentions to raise efficiency and productivity and extant outcomes for resource performance and distributions, and the communities of users dependent on them.

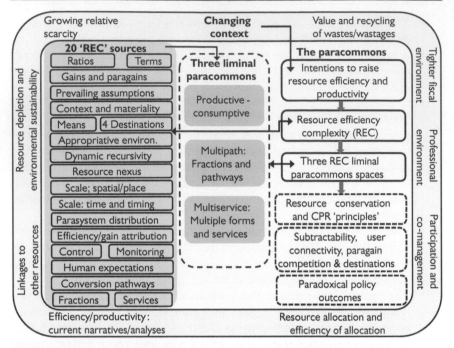

Figure 9 Framework of resource efficiency complexity and the paracommons

2.8 Scope and limitations of the book

The book specifically deals with, and is informed by, the technical efficiency and productivity of natural resources, reflecting the experience and training of the author in soils, irrigation and water. Furthermore the book's ideas are constructed to argue a) that resource use efficiency is far from linear and predictable; b) in a resource-scarce context, complexity resides within the resource as well as from additional layers of the economics, institutions and governance arrangements of resources; c) that common pool conceptions can be applied to the losses, wastes and wastages of resource use, and that; d) users become interconnected via resource use efficiency when distributions of salvaged losses travel to or are held within different systems and jurisdictions.

The book concentrates particularly on the material dimensions of physical resource efficiency – in the conversion in the top part of Figure 10. This shows the efficiency ratio explaining the movement of natural capital from the denominator to create goods and services in the numerator. The goods and services are mainly understood by their quantitative consumption of the physical resource in question (in the case of irrigation as crop transpiration). This physical materiality is, I argue, sufficiently complex as to warrant its investigation as revealed in these pages. However, this is in contrast to the two other ways that efficiency might be viewed, also given in Figure 10. The middle

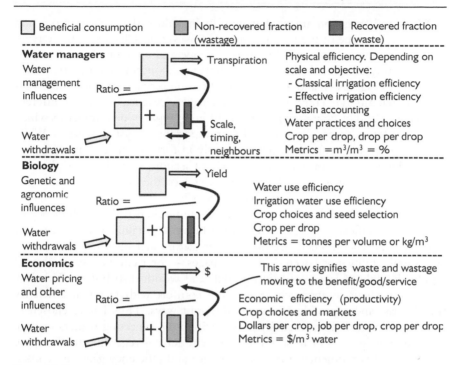

Figure 10 Economic, biological and water (physical) concerns with efficiency

ratio is where the goods and services are biophysical or as 'crop per drop' and is how a biologist might see water use efficiency. The lower ratio in Figure 10 is where the numerator is seen in economic and monetary terms, as job or dollar per drop and might be how an economist sees efficiency. These divisions are a somewhat artificial – in effect they cross refer and inform.

In taking this direction (the top of Figure 10), the author has purposively taken the decision not to approach the topic of efficiency as an economist, biologist, irrigation engineer, sociologist or systems scientist might.[8] Therefore although this book might inform a variety of debates, it does not address the topics of market failure in resource governance; the pricing of resources; game theory applied to collective choice, crop breeding and genetics, crop agronomy and psychological and sociological reasoning of why individuals and communities might consume far less or more resources and generate much lower or higher losses than others in similar circumstances (Frisvold and Deva, 2012). With regards to personal behaviours and views around efficiency (Princen, 2005; Graymore and Wallis, 2010; Hess, 2010) there is not the room to devote to this fascinating dimension implied by US President Jimmy Carter; 'When people see the threat of a future shortage, will increase their wastefulness to be sure they get their share of water that might be scarce in the future that is not scarce today' (c. 1977).[9]

Water and irrigation are used mainly to introduce the paracommons and to reveal a key type of the paracommons. However, in Chapter 5, three types of paracommons are posited. In terms of scope and content, there is not the space to fully address all three types. However, alongside Figure 10, I draw attention to Figures 40 and 41 ahead, on the multiservice paracommons. This explains that different emphases for ecosystem services turn on and off different numerators, and by consequence, create different types of denominators which contain different types of losses – or lesser important ecosystem services. While ecosystem services work contemporaneously in forests, one can also perceive them as being in competition with each other via choices over land use planning or conservation management. Further discussion on the significance of this type of paracommons is contained in Chapters 5 to 7.

In terms of scope, I do not aim to write comprehensively on efficiency or to argue that efficiency is exclusively at the heart of sustainability science – see also Jollands (2006). There are many other factors that inform green growth and sustainability not covered in these pages – for example, issues around equity, the rule of law, population growth and wealth distribution. This lack of a comprehensive approach reflects my comfort with fact that efficiency may be cast and approached in many different ways. Thus I have chosen to frame efficiency and productivity in a specific way, expecting readers to have prior knowledge of – or to seek out – current debates regarding resource use efficiency. My argument is that when savings and efficiency gains are chosen as strategies for reducing resource consumption, the presence of 'efficiency' suggests transition and problematizes the prediction of the outcomes of resource management and sustainability – hence my selection of the term 'paracommons' and examination of contrasts between it and the 'principal commons'. To repeat, the liminal paracommons signals great ambiguity and complexity regarding how to deliver sustainability via efficiency savings. Recognising the early stages of the development of this idea, the book is largely conceptual, aiming to add new thinking to the evolving literature on the commons and natural resources sustainability.

Readers will see the book does not, regrettably, utilise empirical data to support its main arguments, instead employing worked examples. This decision flows from the author's experience and conviction that detailed studies of natural resource systems accommodating scalar levels over long time frames are missing or not up to a satisfactory standard.[10] In irrigation, this gap comes from two key sources. Irrigation donors and research funders have disengaged with long-term expensive irrigation research. Second, engineers and scientists too often report performance findings that are the result of studies at one level (crop, field; system or basin) rather than incorporating all levels. Related to this, methods are not standardised (van der Kooij et al., 2013) or make assumptions (for example, that open canals lose water) rather than employ real tests.[11] The topic of methods in irrigation efficiency studies is picked up elsewhere in the book, under the heading of efficiency kinetics. Although some terms and definitions are outlined at the start of the book and elsewhere in places, the book does not aim to precisely ascribe meaning and specifications to terms used. It has refrained from doing so because it seeks not to constrain a future evolution of terms.

2.9 Chapter review

Figure 11 summarises the chapter. It depicts the role of losses, wastes and wastages in the science and politics of managing the commons, and of the contrasts between expectations of savings via efficiencies for eventual reductions in consumption of natural capital. The left-hand pie-chart shows options for water withdrawals from a common pool (total pie) comprising beneficial and non-beneficial consumption, recovered and non-recovered 'losses' and a fraction not withdrawn. This system is then 'reworked' to boost efficiency and productivity and create a fraction supposedly available for beneficial use by the same user or kept in the common pool for others. Yet outcomes differ from intentions; the exploded pie-chart on the right-hand side of Figure 11 shows the doubts surrounding resources 'freed up' from savings from, and reductions of, loss/waste/wastage fractions. For example, the 'saved' resource might not be measured or traced. It might require significant expenditure to translocate to

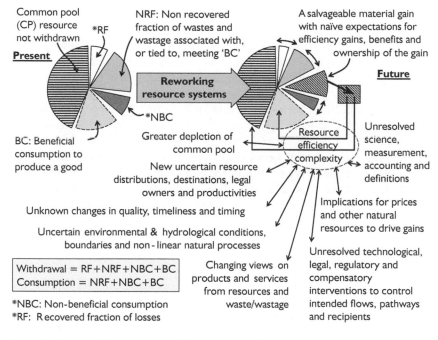

Figure 11 Summarising resource efficiency complexity

other users. It might be valuable in quantity terms but travel too slowly through the river basin to be of timely productive use, or it might result in higher aggregate depletion of the common pool.

The substantiating concerns informing 'resource efficiency complexity' (REC), the paracommons and implications for new approaches are reviewed here, with rest of the book providing exposition.

1 At the heart of resource efficiency complexity and the paracommons is a conversion process of resource inputs to outputs giving rise to losses and to an efficiency or productivity calculation. This ratio hides information and is employed in different ways by different scientific and political constituencies.

2 In a scarce and increasingly 'closed' system world, previously low value or nuisance losses (wastes and wastages) become more valuable. This ratchets up the significance of resource performance, resource cycling/cyclicality and the avoidance of waste/wastage. For example, carbon dioxide, once an externalised wastage of countless natural and human metabolic processes is now increasingly internalised, priced and 'avoided'. In irrigation, volumes of 'wasted' water are large and valuable enough to be the object of further reuse.

3 By being concerned with efficiency, systems and their users are implicitly and explicitly interested in transiting from a current inefficient situation (though often perceived and rarely measured) to a future more efficient situation (also rarely measured).

4 Society faces questions over the means to achieve and redistribute resource gains by salvaging losses, wastes and wastages. This idea is captured in the phrase 'efficiency as resource' and by viewing this material gain as a 'paragain'.

5 The intersection between current and future resource distributions, efficiencies and gains speaks to the heart of the paracommons. New efficiencies change those patterns, altering ownership claims. This generates uncertainty because current and future waste and wastage fractions are often poorly monitored and understood.

6 With reference to increasingly closed systems – for example, planetary or river basin, (Rockström et al., 2009; Falkenmark and Molden, 2008), the connections between scarcity, resource cyclicality and efficiency gains, which combined with changing scales and boundaries, impose new challenges for management and accounting as we seek to understand the additional complexity of new internalities of once ejected waste externalities. This need for accurate accounting and terminology is behind the Jevons Paradox and debates on irrigation efficiency. Likewise, carbon dioxide now vested with value via carbon markets is the subject of budgeting and accounting debates to determine the theory and grounds for judging long-term sequestration (Law and Harmon, 2011).

7 Society is increasingly interested in different forms, localities and qualities of key ecosystem services (such as biodiversity and long-term sequestered carbon) and, as a consequence but depending on circumstances, subordinated or lesser important outcomes (exemplified by rapid turnover carbon).

8 Doubts about the technological, institutional and financial means of raising efficiency shape uncertain outcomes. In irrigation, it is not clear what interventions raise efficiency and crop productivity in a reliable, cost-effective manner while reducing hydrological impact. Furthermore, the attribution and location of savings (e.g. whether they sit with system managers or farmers) raises questions over ownership and cost attribution. The inevitable political promises to deal with these options are part of REC. How these outcomes play out in terms of an equitable and just distribution of post-saving wastes/wastages is also of interest to social scientists.

9 Property claims over resources destined to be lost or saved from one user (for example, seepage from canals) are subject to speculation. Resource efficiency complexity and the paracommons reveal vexed questions of ownership over the fraction of the resource not yet wasted, as well as the resource wasted, plus the resource subsequently 'salvaged' if an efficiency programme is implemented and veritably creates 'real' savings.

10 How these uncertainties consequently (often counter-intuitively) determine the eventual consumption of natural capital is the defining problem of the paracommons. The paracommons forms a heuristic test of the purposes of making 'savings', for example, either for a reduction of total aggregate abstraction and consumption, or allocation to other users, or as an implicit, even conspiratorial, actuator for increased resource consumption.

11 A variety of ontological and epistemological concerns regarding social-ecological systems is illuminated by resource efficiency, for example: the construction of theories of resource use efficiency; the information loss of efficiency ratios; efficiency definition and interpretation differences between disciplines, and scale, level and boundary choices for systems identification. How we perceive social-ecological systems is arguably a reflection of how we perceive resource use efficiency and productivity.

In summary I argue that for some resources a threshold uncertainty space arises out of, and recursively shapes, the science and purpose of resource efficiency and productivity. This space is where the potentials of size, types and destination of resources and their wastage fractions resolve or decompose into different outcomes. The contrast between (usually) optimistic expectations of resource savings/productivity gains set against (often disappointing) consumption and efficiency outcomes produces a political and scientific sphere in which efficiency-change interventions are problematically promised. It is this 'twixt and tween' or liminality that characterises the paracommons and 'resource efficiency complexity' (REC).

Chapter 3

On resource efficiency

Multiple views

This chapter explores some key ideas surrounding efficiency. This non-exhaustive review discusses component parts of the topic while the next chapter employs a more focussed use of literature to present the 20 factors that underpin resource efficiency complexity (REC). This chapter allows efficiency to be seen in a number of ways – for example, as a performance and diagnostic measure, or as a design step. The various subsections of this chapter support the argument that these interpretations are part and parcel of efficiency's complexity, some re-emerging elsewhere in the REC and paracommons framework in ensuing chapters.

3.1 Efficiency in irrigation: challenging systems

A review of some of the ideas and debates surrounding irrigation efficiency in irrigation systems is a precursor to both the remainder of this chapter and to understanding the rationale for taking a complexity view of efficiency.

3.1.1 Complex water distribution and efficiency of irrigation systems

It would be possible to write a different book on irrigation efficiency without referring to the scales, levels, fractions, farmers' views or gains. Possibly 'old school', such a book would not be mundane and without merit. It is the technical difficulty of managing water on irrigation systems that gives rise to substantial efficiency interests without invoking additional complexity brought by nestedness, stakeholder views and so on. It would be a mistake to think that the challenge of irrigation water management exists only because irrigation sits in river basins alongside other demands for water.

There are two main reasons why water distribution in irrigation systems is difficult. The first relates to an ever-shifting mismatch between supply and demand over time and space. Figure 12 pictorially hints at the layers of properties within irrigation systems that lead to this mismatch. At the bottom of Figure 12, the uniformity and magnitude of evaporative demand is given. Potential

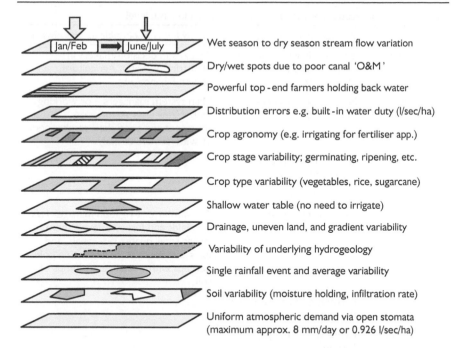

Figure 12 Layers of complexity in achieving efficient and equitable irrigation

evapotranspiration rates are established by climate variables such as temperature, humidity, sunshine hours and windspeed and if we assume irrigation in warm to hot semi-arid locales, then peak requirements can reach more than 7 or 8 millimetres per day.[1] At the scale of irrigation systems and sub-catchments these weather/climate variables tend to be quite uniform unlike rainfall which can be highly variable in amount, intensity, timing and place.

However, moving up Figure 12, multiple factors intervene to make the job of matching a variable water supply to variable demand extremely difficult with the consequence that fields and patches of land are commonly under-irrigated while others are over-irrigated. A changing demand is created by a complex mix of soil type, moisture content, crop type and stage, and farmer influence on distribution. A variable supply is influenced by rainfall, changing streamflow (or groundwater) and internal errors with design, operation and maintenance.

The second reason water distribution is difficult is because river and irrigation systems divide flows of water at smaller and smaller scales using division points such as turnouts and gates. Table 6 explains this fractal (nested) nature of water bifurcation. Within each river and irrigation system, water is drawn from an abstraction point and divided down to the level of plant stomata. To be clear, while a water flow is easy to divide, it is difficult to do so in a way that matches the water demands of the crops and command areas at the lowest level of the system. The task of getting the ratio of 'flow to area' so that the supply

Table 6 Water bifurcations at the crop, field, system and basin level

Level	Water dividing between	Who/what is responsible for this division
River system	Between irrigation intakes; between irrigation and other sectors	Policy- and decision-makers in Ministries of Water, Agriculture, Power, Finance, etc.
Main canal	Secondary canal gates on main canal	Main system managers – or working with Water User Association representatives and their gatekeepers or with farmers directly.
Secondary	Tertiary canal gates and command areas	
Tertiary	Farms or tertiary outlets	
Farm	Fields within the farm	The farmer or irrigation manager lays out fields and furrows, levels them, cleans small quaternary canals, and fixes and operates small division points perhaps with a spade and clumps of soil.
Field	Furrows and zones within the field	
Bunded plot or furrow	Crops within the plot or furrow	
Crop plant	Branches and plant organs in the crop	The crop's biology ensures water drawn up by roots is partitioned to different zones within the crop or plant.
Crop branch	Leaves along a crop stem or branch	
Leaf	Stomata within the leaf	

hydromodule (water duty in litres per second per hectare (l/sec/ha)) matches the demand hydromodule accurately is often not well-engineered into large canal systems.[2] When married to the ever-shifting layers of mismatch complexity in Figure 12, this bifurcation challenge becomes even more difficult to get right. To be completely '100% efficient' would require water managers to precisely divide the abstracted volume at the river in amounts that match the final division at the crop root, crop leaf and stomatal level.

The outcomes of matching a variable supply with demand can be seen in Figure 13 (also Borgia et al., 2012). The left-hand side, taken in Tanzania, shows a highly variable pattern of water and crop growth in rice irrigation. The right-hand photo, taken in Swaziland, shows a much more uniform pattern of growth seen in sugarcane cultivation on private estates. The reasons for the differences are manifold, but in simple terms, despite both being mono-crop estates on heavy clay soils, the rice growers in the Tanzanian public irrigation scheme

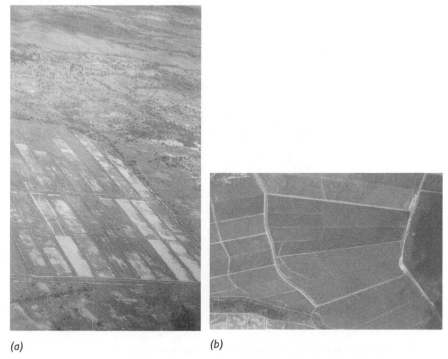

(a)   *(b)*

Figure 13 Comparing variable and uniform growth patterns on irrigation systems

Photos show (a) variable wetting and growth patterns on rice irrigation systems in Southern Tanzania
and (b) more uniform growth patterns seen in sugarcane irrigation systems in Swaziland

were less concerned about water distribution. Yet while the sugarcane growers
look to have 'solved' the problem of mismatching supply to demand, this photo
(showing about 40 hectares of cane) is part of much larger system (of about 5000
hectares) where the total problem of managing water remains considerable,
particularly after rainfall events, or to accommodate droughts, harvesting and
crop growth variability.

It is worth summarising how water management complexity within irrigation
systems has significance for resource efficiency complexity and debates around
efficiency. This within-system complexity points to the need to understand the
details of water management at the spatial scale of the field and farmer and at the
time scale of hours and half-days.

- River basins should be seen as made up from many fields, farmers and crops.
  This counter-balances the view of the river basin as a monolithic block
  comprising one inflow and three or four outflow fractions. Furthermore,
  the layers given in Figure 12 do not simply and neatly scale up – instead the
  layers replicate unevenly and incoherently over time and space.

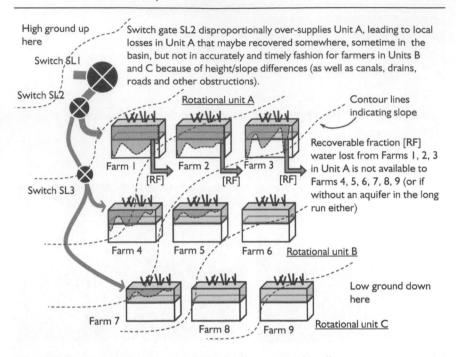

High ground up here

Switch SL1

Switch SL2

Switch SL3

Switch gate SL2 disproportionally over-supplies Unit A, leading to local losses in Unit A that maybe recovered somewhere, sometime in the basin, but not in accurately and timely fashion for farmers in Units B and C because of height/slope differences (as well as canals, drains, roads and other obstructions).

Rotational unit A

Farm 1   Farm 2   Farm 3
[RF]   [RF]   [RF]

Contour lines indicating slope

Recoverable fraction [RF] water lost from Farms 1, 2, 3 in Unit A is not available to Farms 4, 5, 6, 7, 8, 9 (or if without an aquifer in the long run either)

Farm 4   Farm 5   Farm 6   Rotational unit B

Low ground down here

Farm 7   Farm 8   Farm 9   Rotational unit C

Figure 14 Recoverable flows are not immediately recovered to all

- Excess irrigation water in one location is not picked up by other under-irrigated farmers except in particular circumstances. The reason farmers cannot quickly and easily compensate for over-irrigation elsewhere is because of the fall of the land, distance and water current velocities involved. In other words, farmers served by a single canal are unlikely to be able to transfer each other's water recoverable losses. Figure 14 shows how water can only be recovered in certain circumstances. Farmers can pick up drainage water when they sit directly downstream on the same drainage line. Even with this, the recipient farmer can only use this drainage water if she has command (the water level is above ground level) or uses a lifting device.
- Furthermore, there will be delays while water first wets up the soil profile and then moves below the root zone or to the end of the field. Water takes time to move through fields and soil – far more slowly than when flowing through river and drainage channels. My observation is that water moves in channels from rates of about 0.2 to 2 metres per second (so that 2000 metres is covered in about one hour). Water movement through soil and aquifers is in the order of between 1 to 200 centimetres per hour; a distance of 2000 metres takes about 42 days to cover. This delay may or may not be a problem depending on a particular mix of production cycles within or between seasons. For example, in southern Tanzania, slow moving water

means that drainage-reliant farmers start their irrigation later in the wet season with the result that the rice ripens in the cooler temperatures of April and May which reduces yields by about two-thirds (Machibya, 2003).

3.1.2 A synopsis of the efficiency debate in irrigation

There are many debates in irrigation – for example, on irrigation expansion in Africa, sustainable groundwater abstraction in Asia, and the transfer of ownership of large systems to smallholders. While these intersect with productivity and efficiency, the topic of efficiency in irrigation is, in its own right, subject to scrutiny. These pages contain multiple references to this debate – for clarity I provide a synopsis here. This synopsis is for the benefit of readers whose knowledge of debate is limited; there are many other side-issues not included here.

There are essentially three schools of thought in irrigation efficiency: fractions, effective efficiency and classical efficiency (though in Lankford (2012b) I joined the former two together).[3] Figure 15 contains the three ideas. Fractions are 'portions' of the withdrawn volume of water, divided into beneficial fraction, non-beneficial fraction, recovered fraction and non-recovered fraction. Effective irrigation efficiency (EIE) is the computation of beneficial consumption over the total consumed amount. Classical irrigation efficiency (CIE) is the computation of beneficial consumption divided by the total withdrawn amount.

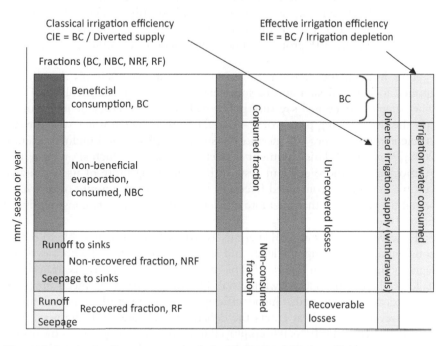

Figure 15 Accounting for recovery: classical and effective irrigation efficiency

One might ask how the accounting in Figure 15 leads to contentious debate as seen in recent issues of the journal 'Water International' (JWI). While I risk opprobrium from various protagonists, these are my interpretations of this question:

- Fractions are favoured for *not* incorporating a ratio calculation of efficiency because efficiency hides information about the total amount of water consumed in basins. One purpose of fractions is to highlight the risk in river basins of excessive total consumed volume. However, this entirely worthy argument is often not fully understood by irrigation managers who continue to talk in terms of 'efficiency'.
- Effective efficiency computes a performance ratio allowing basin managers to understand irrigation systems from the basin's perspective. Thus a system that is 83 per cent EIE puts 83 units through beneficial consumption for every 100 units of total consumption. However, by using a ratio, EIE fails to depict total consumed volumes (not satisfying the fractions camp) and does not allow managers to understand the efficiency of their own system (not satisfying the CIE school).
- Classical efficiency allows irrigation managers to discern the performance of their particular system of water use. Thus a CIE of 64 per cent tells a manager that 64 units of every 100 units withdrawn from a river go to beneficial consumption. However, this performance measure does not depict total consumed volumes in a basin (dissatisfying the fractions group) or that part of the water which returns to the river basin (not satisfying the EIE group, or indeed the fractions school too).

The 'heat' in the debate revolves around how these schools thought are applied to policy design and to some extent on the science involved. Those who support 'fractions' say that policy calls for more efficient irrigation to run the risk of being paradoxical – that far from reducing water withdrawals, efficiency programmes will increase consumption (Ward and Pulido-Velázquez, 2008; Perry, 2007) This argument is given in Figure 16 which shows a system 'A' moving to a more efficient system 'B', converting more of its withdrawn volume of water to consumed fractions. By lowering the amount of recovered water, the paradox is that the more efficient system 'B' is more consumptive of natural capital.

However, Gleick *et al.* (2011) argue, as would I, that it is possible to reduce consumption by raising efficiency and simultaneously altering withdrawals via legislation, as shown in Figure 17. The more efficient system 'B' has transferred its lower 'losses' into a lower withdrawal. This means system 'B' meets the (non-paradoxical) expectation that higher efficiency is less impactful on natural capital.

However, further heat enters the debate because the entire framework of 'fractions, EIE and CIE' is not employed to its full advantage which means the various schools appear (to me) to be missing the advantages of each other's

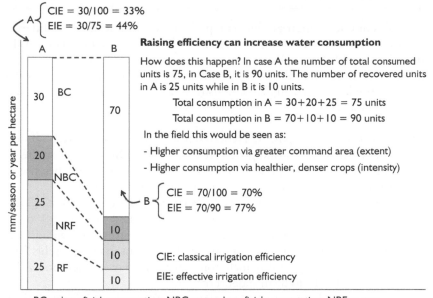

$A \begin{cases} \text{CIE} = 30/100 = 33\% \\ \text{EIE} = 30/75 = 44\% \end{cases}$

Raising efficiency can increase water consumption

How does this happen? In case A the number of total consumed units is 75, in Case B, it is 90 units. The number of recovered units in A is 25 units while in B it is 10 units.

Total consumption in A = 30+20+25 = 75 units

Total consumption in B = 70+10+10 = 90 units

In the field this would be seen as:

- Higher consumption via greater command area (extent)
- Higher consumption via healthier, denser crops (intensity)

$B \begin{cases} \text{CIE} = 70/100 = 70\% \\ \text{EIE} = 70/90 = 77\% \end{cases}$

CIE: classical irrigation efficiency

EIE: effective irrigation efficiency

BC = beneficial consumption, NBC = non-beneficial consumption, NRF = non-recoverable fraction, RF = recoverable fraction.

Figure 16 Raising efficiency can paradoxically raise consumption

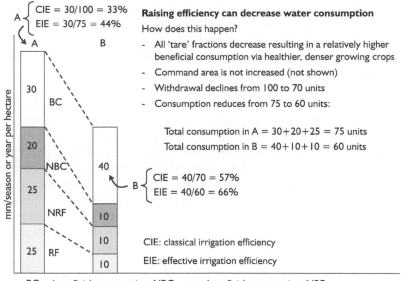

$A \begin{cases} \text{CIE} = 30/100 = 33\% \\ \text{EIE} = 30/75 = 44\% \end{cases}$

Raising efficiency can decrease water consumption

How does this happen?

- All 'tare' fractions decrease resulting in a relatively higher beneficial consumption via healthier, denser growing crops
- Command area is not increased (not shown)
- Withdrawal declines from 100 to 70 units
- Consumption reduces from 75 to 60 units:

Total consumption in A = 30+20+25 = 75 units

Total consumption in B = 40+10+10 = 60 units

$B \begin{cases} \text{CIE} = 40/70 = 57\% \\ \text{EIE} = 40/60 = 66\% \end{cases}$

CIE: classical irrigation efficiency

EIE: effective irrigation efficiency

BC = beneficial consumption, NBC = non-beneficial consumption, NRF = non-recoverable fraction, RF = recoverable fraction.

Figure 17 Raising efficiency can follow expectations in reducing consumption

points. A common language and comprehensive classification of different approaches to water accounting still seems to be missing (and is beyond the scope of this book). Although complex issues are mentioned in the JWI exchange (and complexity is referred to), complexity as a holistic frame for efficiency is poorly captured. A few examples demonstrate on-going questions and possible misunderstandings:

- In the latest development, Frederiksen and Allen (2011) have omitted the non-beneficial consumptive (NBC) fraction. This decision means all irrigation systems are precluded from having reducible losses because all fractions are either consumed, or recovered or not recovered.
- The previous point is emphasised by Perry in the Frederiksen *et al.* (2012) response when he says 'there is no reference to "new" water' (p.196). In other words, by accounting for each and every portion of a withdrawal, new water cannot be inserted. However, Perry reinserts NBC in his 2007 framework on page 195 which contradicts the Frederiksen and Allen position he seems to be in support of.
- Lankford (2012b) points out how timing is affected by efficiency – which the JWI authors in the debate have not fully accommodated. Frederiksen and Allen, 2011 on page 275, use an example 'with no timing effects'.[4]
- Gleick *et al.* (2011) following on from PI (2010) undermine their own argument by a) using terms and amounts that offers too much certainty ('new water' and 'million acre-feet') while proposing technologies that typically increase the consumptive use of a proprietor's water from river basins such as switching to sprinkler and drip irrigation – a technology that also requires maintenance and reliable and cheap power to run.[5]
- Lankford (2012) argues that fractions, EIE and CIE each have their value for acting on water management in river basins. A classification of 'fit' for purpose also seems not to interest the JWI debate.

What is also worrying, though understandable, is that these discussions are conducted with minimal reference to trustworthy empirical data at both the field and basin levels. Frederiken and Allen appear to be arguing that non-beneficial consumption (seen on the right-hand side of Figure 8) is negligible to an extent that their computation does not allow it (not even for systems where it might be present). With this degree of miscommunication (contra to Perry's 2007 objective of improving communication), it is entirely predictable that the paracommons' treatment of efficiency is seen by 'CIE, EIE and fractions' protagonists as unhelpful in the cause of their particular viewpoint.

3.1.3 Simplifying and omitting efficiency

The treatment of efficiency-type conversions in the conservation of resources is problematic. Certainly I can testify to a personal sense of unease when I hear the everyday use of the term 'efficiency' applied to resource management in the public realm. It seems to invariably denote a costless and simplified 'outcome' that would solve a range of societal and environmental imbalances.

I believe that efficiency has been marginalised to the extent that it has been simplified or omitted altogether. These two types of usage (simplification and omission) are very distant from the exposition of efficiency complexity contained in this book. While a philosophical discussion of society's relationship to efficiency is beyond the remit of this book (for example, Makoff, 2011; Princen, 2005) I shall instead record two 'omission' manifestations regarding the terms and science of resource use efficiency.

The first is a passive omission of the topic in literatures on natural resource management – mostly explained by a supplanting of efficiency by productivity (see Section 3.2 for more on this). Despite early discussions of the links between conservation and efficiency (Hays, 1959)[6] plus a robust debate on the Jevons Paradox on energy efficiency and emerging literatures on eco-efficiency and Factor X ideas, many mainstream natural resource, commons and ecosystem texts in the last 20 years have paid little attention to responses to conservation and resource sustainability via efficiency improvements. To mention a few, Leach et al. (2010), Homer-Dixon (2001) and Adams (2008) treat conservation and distribution via the capping of total consumption to match supply. Whether and how consumption is reduced by managing waste/wastage or by treating withdrawal and consumption differently is rarely or insufficiently discussed. Texts on agricultural systems also treat efficiency very lightly or miss it altogether (e.g. Snap and Pound, 2008) or wrap it up as productivity (eco-efficiency; CIAT, 2012).

One result of an emphasis on productivity is that the building blocks and mechanisms of the consumption of natural capital – and the consequences of making savings – are poorly defined. This omission is paralleled perhaps more surprisingly in the irrigated agriculture literature, with the exception of specialist articles such as the Comprehensive Assessment of Water Management in Agriculture (CAWMA, 2007) (yet which emphasised productivity over efficiency). Many mainstream texts on the management and conservation of water and river basins in the face of scarcity, including agricultural water, do not unpack efficiency and productivity (see, for example, Falkenmark and Rockström, 2004; Svendsen, 2006; Pearce, 2007; Barlow, 2009; Lenton and Muller, 2009; World Water Assessment Programme, 2009; Rogers and Leal, 2010; Solomon, 2010; The Economist, 2010; Chartres and Varma, 2010; Matthews et al., 2011). This omission in this list is surprising given the profile offered to efficiency within the irrigation engineering literature and to potential-but-unverified gains to be had from even mediocre increases in irrigation efficiency as indicated in the second chapter.

For example, the Chartres and Varma's (2010) treatise on solving the world's water problems, which points to the complexity of irrigation efficiency (p. 193), avoids any examination of its theoretical meaning and is relatively light on solutions to improve efficiency. The chapters in Svendsen's 2006 book also stay clear of any mention of how an improvement in efficiency might help resolve the challenges of closing river basins (more is made of productivity). At the recent 2012 Stockholm World Water Week, the presenters from IWMI, FAO and ICID in the session on 'Best Use of Blue Water Resources for Food Security' kept clear of efficiency as a topic, but did consider productivity.[7] Furthermore the lack of an entry on Wikipedia.org (visited 27 August 2012) on irrigation efficiency provides me with further evidence of its minimal profile. Wikipedia is not a rigorous measure of scientific understanding but more a gauge of societal concern; within Wikipedia's compendium where a great number of esoteric and mainstream subjects are accessible, other environmental discourses are discussed in detail. Wikipedia includes entries on: deforestation, desertification, acid rain, soil erosion, biodiversity, environmental degradation, and the water crisis.

In addition to passively omitting efficiency, it is possible to argue that some organisations became sufficiently concerned about the term 'efficiency' to take the decision to actively refrain from using the term. During the last 10 to 15 years, various irrigation organisations have discussed the decision to cease using the term – these include IWMI, ICID and the Journal Irrigation and Drainage – and move towards accounts for final water environmental outcomes at the basin level.[8] The quote below is in Keller and Keller (1995, p. 18 fn. 2) reflects some of the conversations regarding efficiency I have had with others resulting from my London 2008 organisation of a seminar on efficiency and productivity described below.

> Subsequent to our first draft of this paper, we discovered that R. G. Allen and L. S. Willardson have been writing a paper, Elimination of irrigation efficiencies (to be presented at the 13th Technical Conference on Irrigation, Drainage and Flood Control, Sponsored by the U.S. Committee on Irrigation and Drainage, Denver, Colorado, October 19–22, 1994), pointing out that what we call classical irrigation efficiency is an outmoded term. They suggest that irrigationists stop using irrigation efficiency terms because the use "is interfering with rational management and planning of the use and allocation of water resources." They recommend the use of ratios or fractions instead because classical irrigation efficiency terms, which do not consider reuse of return flows, have been misapplied so often (as pointed out by Jensen, 1977). In their presentation of the use of ratios in place of efficiencies, they consider the degradation of water due to salt build up and pollution, and consider the need for leaching. Although we do not necessarily agree with eliminating the use of the efficiency concept, we applaud their crusade to correct the misunderstanding of classical irrigation efficiency terms.

The provenance and purpose of the 2012 *Agricultural Water Management* special issue on irrigation efficiency and productivity stemmed from a seminar entitled 'Towards a political ecology of irrigation and water use efficiency and productivity' held at the Institute for Civil Engineers in London on 6 November 2008. This was arranged by the author on behalf of the British Section of ICID (International Commission on Irrigation and Drainage). The meeting intended to showcase the theories and understandings accompanying emerging debates on water productivity and efficiency. Although the programme's intention was pluralistic and open to interpretation, one reason for calling the meeting arose out of my concern that 'irrigation efficiency' as a term, and as a means to reflect deeply on water management, was being unnecessarily marginalised (Willardson *et al.*, 1994; Seckler, 1996; Perry, 2007; Jensen, 2007). These papers, having drawn attention to flaws in applying irrigation efficiency in the context of water allocation, went on to suggest that irrigation efficiency had limited utility for irrigation management and should be dropped from usage. If efficiency was to be employed, it would be utilised in its classic sense for irrigation design.

The marginalisation of the term efficiency is understandable; it comes from a widespread failure to discern how efficiency plays out in natural resource management and policy. Policies that call for higher efficiency invariably look circumspect when depending wholly on narrow but dramatic technological change, for example, the adoption of drip or sprinkler systems (PI, 2010; 2030 Water Group, 2009). In other words, 'efficient systems' are not necessarily the same as 'non-consumptive systems' because the former might drive up consumption of natural capital and the latter implies a cap on consumption. Efficiency, as a way of understanding systems, needs to explore the many dimensions and consequences of efficiency without giving oxygen to an uncritical naturalising policy (Boelens and Vos, 2012). By assuming 'efficient' is an unalloyed benefit, science and society fail to recognise the disbenefits that efficiency policy can bring.

3.1.4 The evolution of irrigation efficiency

A number of authors have traced the evolution of the terms and definitions of irrigation efficiency (e.g. van Halsema and Vincent, 2012); Table 7 gives a summary of this history. Several observations may be made; first, this evolution reflects an increasing understanding of the complexity of irrigated agriculture. Second, these definitions are not replaceable paradigms; none of the computations are technically 'wrong', rather they have their own applications for specific conditions (I do, however, believe that the field methods for arriving at their calculation are often flawed – a discussion picked up in Section 7.3). Therefore it is the appropriateness of each measure that is the significant issue here. Third, it is highly likely that this evolution has not stopped. We are entering a much more scrutinised era regarding irrigation efficiency and 'freed up resources'; one that will demand new thinking and metrics.

Table 7 Evolution of irrigation efficiency terms at the system and basin level

Term	Computation	Authors	Notes
Irrigation efficiency	Crop transpiration divided by total withdrawn water	Israelson, 1950; Jensen, 1967; Jensen, 1983; Bhuiyan, 1982; ICID, 1978	Now seen as 'classical irrigation efficiency' (CIE) (with further allowances for rainfall, salinity control and crop cooling).
Application efficiency	Multiple measures exist designed to explore uniformity and adequacy of irrigation at the field level	Jensen, 1983	Often irrigation engineers combine conveyance efficiency with field level application efficiency to arrive at system efficiency. This, I feel, is erroneous and misses how water moves within irrigation systems.
Conveyance efficiency	Ratio of water delivered to users by water withdrawn from sources	Bos, 1979	
Effective irrigation efficiency	Crop transpiration divided by total consumed water	Keller and Keller, 1995; Seckler, 1996; Haie and Keller, 2008	Boundaries of system tend to be at the basin level.
Fractions	Volumes of water as final outcomes having passed through irrigation systems	Willardson *et al.* 1994; Perry, 2007	Ratios are generally not employed.
Irrigation sagacity	Water consumed reasonably divided by total consumed water	Solomon and Burt, 1999	
Attainable irrigation efficiency	Water consumed by water-short irrigators divided by water consumed by water profligate users	Lankford, 2006	A local, relative framing of what is attainable by referring to careful irrigators within the vicinity.

3.1.5 Efficiency as a design step

By referring to the debate on efficiency within the discipline of irrigation management, it is possible to discern how the dimensionless measure of irrigation performance is employed to design irrigation systems. Thus in order to deliver 'x' litres per second for the crop, an efficiency correction factor has to be applied in order to abstract 'y' litres per second at the river or aquifer in order that 'x' l/sec is correctly delivered following a sequence of losses. While mathematically and technically this presents no problem, the procedure is problematic because the efficiency figure employed is often over-estimated and not further tested and refined – leading to over-abstraction for the system to the detriment of shares and sharing with other systems and users within the river basin (see Lankford, 2012b; Sections 4.6.4 and 6.1.5 of this book for more discussion). It is also problematic because this design procedure can be mistakenly employed when subsequently characterising the actual efficiency of the system post-construction. Finally, a key confusion arises because although this procedure as a design step, it is not at all clear that irrigation practitioners and designers go on to distinguish between classical and effective forms (Molden *et al.*, 2010).

3.1.6 Efficiency as design

In several papers in the recent *Agricultural Water Management* Special Issue (2012), Boelens and Vos, Lankford, Lopez-Gunn *et al.*, van Halsema and Vincent observe that national policies to improve the efficiency and productivity of irrigation depart from a naïve technological characterisation of traditional systems in decline, suffering from low irrigation efficiency. Such policies are defined by modernisation rhetoric in three ways. First and commonplace is the idea that drip or sprinkler systems are more efficient than canal or bucket technologies; the term 'traditional' is commonplace for describing canal/surface systems and is often a by-word for inefficient/inefficiency.[9] Papers extolling the efficiency virtues of drip (Narayanamoorthy, 2004) are unhelpful if they fail to unpack the risks of additional water consumption, costs to farmers of poor installation design, running and maintenance costs, and energy costs that either fall to the farmer, society as a whole or to global carbon (van der Kooij *et al.*, 2013).

The second naïve characterisation is that the installation of drip and sprinkler systems, meters, computer aids and automatic gate control on canal networks define modern and modernising infrastructure. While these might function in certain environments, there are alternative 'low-tech' iterations that can be made to existing systems (Plusquellec, 1994; Renault *et al.*, 2007). Examples include installing modular gates, raising canal density, ensuing an accurate leadstream flow to command area ratio, adding night storage reservoirs, improving the access to in-field water control, and switching from level open fields to furrows (additionally with surge irrigation or alternative row irrigation) The nuanced technologies and practices of an incremental improvement programme are very different from ideas that favour whole step-changes and packages of sprinklers and drip systems.

Related to the two modernising ideas mentioned, is a third modernising rhetoric: that of training farmers in water management, remaking them as 'efficient users' of water: 'Farmers must be trained on soil and water technologies to enhance crop production and food security' (ASARECA, 2006). As Boelens and Vos (2012) observe, these programmes explicate government intentions towards environmental stewardship through the objectification of efficiency and productivity with little understanding of efficiency's composition or of how farmers come to understand it. The outcome (also recorded by Boelens and Vos, 2012; Lopez-Gunn *et al.*, 2012; Knox *et al.*, 2012) is an inability to partner with farmers to solve either national-level concerns regarding food production or local-level concerns regarding top-tail water sharing. An acknowledgement that irrigation efficiency and productivity are by-products emerging from multiple factors managed by farmers, rather than being inanimate design features, would dramatically change the relationship between irrigators, ministry engineers and other actors – in effect making them servants of irrigators rather than vice versa (this topic is continued in Sections 3.5 and 7.1.4 below).

3.2 Efficiency, sufficiency, productivity and sustainability

In this section, I explore a number of themes regarding productivity and sustainability.

3.2.1 Efficiency, eco-efficiency and productivity

The paracommons concept is chiefly about 'efficiency as resource', the idea that putative material savings arise out of efficiency improvements and are then subject to claims and competition commonly without being open and transparent. However, to understand this emphasis requires physical efficiency to be placed alongside economic and biological efficiency and to explain where the concepts of eco-efficiency and productivity fit in this relationship.

To this end, Table 8 sketches a non-exhaustive classification of different efficiency and productivity ratios moving from dimensionless efficiency measures at the top to complex synoptic dimensional measures in the lower row. This table may be read alongside Figure 10 which shows the three ways that efficiency may be viewed. The top part of Figure 10 and Table 8 pertains to physical dimensionless measures such as efficiency. The bottom two diagrams in Figure 10 relate to biological and economic measures of efficiency. The last row of Table 8 on ecosystem services is addressed in Section 3.4.2.[10]

Table 8 mainly distinguishes between dimensionless and dimensional efficiency measures. Thus, in irrigation systems, irrigation efficiency is a unitless or dimensionless ratio (e.g. 74 per cent) while productivity is a ratio of mixed

Table 8 Conversion ratios within natural resources

Type of conversion	Example	Units
Dimensionless 'efficiency' type	A measure of system management performance. For example: beneficial water consumption to total withdrawals of water.	Cubic metres/cubic metres (expressed as %)
Dimensional efficiency or productivity, performance ratio	A measure of a physical or economic output per input. • Economic • Biological (water use efficiency) • Qualitative	US$/cubic metres of water Kg beef/m³ water or kg sugar/m³ Nutrient load category /m³
Dimensional synoptic production or services	Carbon sequestered as biodiversity or as soil carbon or as ecosystem services or as hydrological stream flow.	Timber tonnage/hectare Harvested forest products/ ha Runoff regime/baseline regime

units of output (e.g. crop biomass in tonnes) to an input, (e.g. water volume in cubic metres). Dimensional efficiencies such as biological or economic productivity, irrigation water use efficiency and water use efficiency should always be defined in terms of the numerator, denominator, boundaries and methods employed (van der Kooij *et al.*, 2013).[11] The term eco-efficiency (CIAT, 2012) is dimensional in a narrow way (boosting crop yields per hectare) but carries an aspirational message of reducing environmental impact.

A number of points can now be made about the framework in Table 8. Both dimensionless and dimensional measures can be used as instantaneous measures of current performance, and in the case of dimensionless percentages as containing implicit information on a future idealised target (implicitly taken as the full 100 per cent).

A dimensionless irrigation efficiency figure is not without 'material' – dimensionless is not material-less. A (classical) irrigation efficiency figure of say 64 per cent means that normalised to 100 units, 64 units are beneficially consumed and 36 units are 'lost' to that zone of production. Equally a change in efficiency, although dimensionless, gives material outcomes. A change from 64 per cent to 75 per cent efficiency means that 11 units that were previously lost to that zone are now being beneficially consumed. Example calculations of ratios and material units are given in Sections 4.5.2 and 5.2.3.

Thus dimensionless efficiency figures, although primarily about performance, are part and parcel of a material gain when systems transit from an assumed current inefficient state to a future (intended) efficient state. Even with this in mind, as current measures of performance, efficiency figures do not convey how much of a material gain in resource can be achieved by a move to a higher efficiency. For example, in moving from an irrigation efficiency of

64 per cent to a new efficiency of, say, 75 per cent does say how much of the 11 per cent 'savings' are divided between recovered losses and non-recovered. (Furthermore I believe few farmers, engineers and policy-makers are fully aware of the assumptions involved in measuring efficiency.)

Table 8 also invokes the question of how these different measures become linked. Past and recent work seems to have delinked efficiency and productivity, creating the possibility for productivity policies to be fashioned accordingly (see, for example, CAWMA, 2007). Such hypotheses ask whether productivity can be raised without raising classical or effective efficiency or whether productivity and efficiency are always and invariably delinked. Related, are the perennially complicated issues surrounding the denominator and whether this utilises consumptive, depletive, non-consumptive or through-flow water, whether these come from green (rainfall) water or blue water – and how they are measured and arrived at. Playán and Mateos (2006), for example, derive five different denominators depending on these variations.

These productivity emphases have part complicated the efficiency/productivity field – and in doing so they have part energised and challenged the efficiency debate in new ways, five of which are discussed here:

First, the focus on productivity from a water scarcity/water allocation perspective has forced a wider view of the world – beyond a technological view of the irrigation system. The introduction of economic and productive efficiency rather than a process/technological formulation has rightly pushed society to investigate institutional and market mechanisms that economically leverage the greatest productivity gains, thereby driving water towards sectors that produce the greatest economic benefit (see van Halsema and Vincent, 2012; Medellín-Azuara et al., 2012). Allied to this are also questions about institutional and transactional efficiency, asking how might allocation itself be promulgated efficiently.

Second, associated with this economic paradigm are further questions about how we technologically solicit new distributions of water allocation. These new distributions consider higher water productivity to be the utilitarian objective (move water out of low value agriculture to high value industry). This recognises that water depleted from different sectors underpins the productivity increase. Therefore associated policy to raise productivity should pay close attention to water accounting (Karimov et al., 2012).

Counter-balancing this view (the third form of connection) is the argument by Lankford (2012 and 2006) that efficiency and productivity are inseparable. contending that efficiency reflects the on-farm/on-system control of water scheduling and timeliness which affects crop growth and productivity. In other words, in basins where irrigation dominates, the technical route or pathway to greater agricultural and economic productivity will comprise, to a large extent, knowledge of irrigation management and efficiency at the field and farmer level.

A fourth, wider, form of the links between productivity and efficiency is hinted at by the work Karimi et al. (2012a). By adding carbon to the equation (or other inputs and 'wastes') we can continue to enrich our understanding of what constitutes productive, efficient water management while meeting wider system and societal sustainability objectives and constraints.

A fifth form takes a cultural and livelihoods wider view – hinted at in the arguments articulated Lopez-Gunn *et al.* (2012) These authors argue that investments in more 'modern' (in this case effective) forms of control across many irrigation farms within a region offer emerging benefits of farming and irrigation flexibility, including an improved lifestyle of farmers no longer dependent on waiting their turn, and greater adeptness and control of supplemental irrigation, also at the farm and basin level, allowing a mixing of blue water to maximise the productivity of both blue and green (rainfall) water. In other words productivity, efficiency and forms of production as a livelihood and culture cannot be disassociated; in some places modernising to drip systems might be highly suitable while in other places, such a rapid technology change might fail (especially after the first few years). To assume one can optimise water efficiency through introduction of drip and sprinkler systems without referring to current trajectories of production and co-production by collectives of farmers is to take a considerable risk.

In summarising, the connections between efficiency (as a unitless ratio) and productivity (as a ratio of an output, e.g. crop biomass, per water volume) form a rich milieu of research and theory-building to say nothing of the need to clarify differences between emerging terms and understandings. The view taken by this book is that with the exception of the non-water means to boost crop production (e.g. fertiliser, seed choice), productivity is closely linked to efficiency; more efficient irrigation systems, *inter alia*, are more productive systems. It is for this reason that, expediently for the paracommons, the term 'efficiency' largely incorporates 'productivity' except when a distinction is required.

3.2.2 Efficiency and sufficiency: supply, demand and share management

Wary of initiating too many contentious issues in this book, I intentionally use this sub-section to consider briefly how efficiency and the saving of water fits into supply management, demand management and share management. Of interest is how physical efficiency forms one of five components of 'demand management' and thus how demand for water from limited freshwater supplies may be kept down while maintaining or boosting production. By so doing, I initially cast the solving of the scarcity/sufficiency problem as a two-way decision (fixing scarcity by either emphasising 'supply management' or 'demand management' – see, for example, Tortajada, 2006). In other words, given a situation when supply is unable to meet demand how might savings and efficiency enable sustainability and 'square the circle'? This two-way decision allows me, in turn for the remainder of the sub-section (and indeed the whole book) to move beyond this dichotomy and recast efficiency also contributing to a third idea of 'share management'.

Taking the previous introductory paragraph more slowly – I propose that resource use efficiency is one of five components of a demand management approach to solving a water scarcity/sufficiency problem – four of which, including efficiency, were introduced in Table 3 above. The five components are therefore a) reducing losses via raising efficiency; b) reducing net demand via dematerialisation; c) reducing net demand via retrenchment; d) reducing net demand by substitution and; e) pricing water so that the cost of the using the resource pushes down its demand (exemplified by rising tariffs in Singapore described by Tortajada, 2006).

Supply management can now be contrasted to demand management; in order to solve scarcity, more supplies of water need to be sourced, built or otherwise obtained. Four main examples of supply management include dam-building, installing boreholes (or deeper boreholes), inter-basin transfers of water (Snaddon et al., 1998; Gupta and van der Zaag, 2008), catching rainwater in small storage bodies (Wisser et al., 2010), and desalinising salt water into freshwater (Tortajada, 2006).

Confusingly, the conjunctive use of several supply solutions together allows users and engineers to invoke the idea of 'savings' even if underlying demand for the resource remains unchanged. In other words, if one supply is boosted, this provides a 'saving' of another supply. Souza and Ghisi (2012) observe that harvesting of rainwater 'saves' water drawn from a formal potable water supply network.[12] This trade-off highlights several issues pertaining to the paracommons. One relates to the meanings of words employed in balancing supply and demand at different levels and scales. The choice of the word 'saving' by the authors infers a volumetric reduction of water drawn from another, more formal, source. However, their paper does not define the meaning of the words 'save' and 'saving'. The resource efficiency complexity framework in Chapter 4 refers to the ambiguities of terms used to explore efficiency and savings. Secondly, quantitatively, additional water supplies fed into households and communities makes it difficult to isolate and attribute the impacts of any efficiency interventions made in other parts of the chain of water supply and demand. Third, Souza and Ghisis assume that the rainwater harvested by households has no other use or claimant. This may apply, given the likely minor withdrawal by rainwater harvesting devices. However, the meaning of the term 'saving' by using conjunctive supplies to meet urban demand is narrow and possibly naïve; such 'savings' do not recognise the scalar and nested dimensions of complex systems sitting alongside neighbouring competing sectors. Taking these last two points together, claims for aggregate 'savings' with consequences made for natural water capital are difficult to verify.

Broadly speaking, supply management saw favour for the most of the twentieth century. Then in the mid-1980s through to about the start of the new millennium a series of concerns led to a swing towards demand management. These concerns could be associated with: anxieties regarding the environmental sustainability and social costs regarding dams (World Commission on Dams,

2000); worries related to overdraught of aquifers, and the increasing trust vested in a neo-liberal market approach to resource management and allocation seen in the Dublin Principle of water as an economic good (Solanes and Gonzalez-Villarreal, 1999).

Since about 2005, two new agendas have arisen making easy categorisation of supply and demand management difficult. One is 'water security', recognising supply augmentation (Grey and Sadoff, 2007) but also incorporating the demand thinking by recognising the need for efficiency and productivity (Clement, 2013). The other is 'green growth' (e.g. OECD, 2011) also accommodating of both sides of the approach – that building the economy (and the size of the pie) is to be welcomed but doing so in an efficient and environmentally-friendly manner.

There is a third school of thought which in some ways has an uneasy relationship with the rather utilitarian dichotomy of supply and demand management, that of share management. There are potentially different interpretations of this, incorporating, for example, distinctions between allocation, appropriation and distribution (Lankford, 2011). However, and more significantly, the same article places an emphasis on how different properties of water (e.g. visible equity and proportions, time, timing, depth, accurate placement equity, quality) as well as quantities of water are managed. It is these less recognised and less tangible properties of water, and how they are used to meet individual and community needs, that make share management complicated – and into which the idea of the management of the inefficient part of water use fits as a property not usually noticed alongside the volumetric emphasis of supply and demand management.

Moreover, the idea of the paracommons fits with all three schools. The paracommons is an extension of efficiency (demand management); it envisages new supplies borne of efficiency gains (supply management), and it ties together users via the re-routing of the material gains to the four destinations of the parasystem (share management).

3.2.3 Efficiency as resource governance

Furthermore, technical (or physical) resource use efficiency can be incorporated into (and indeed confused with) the 'political economy' or governance sense of fitting an appropriate regulatory and administrative regime to a resource that in turn changes patterns of consumption and production – for this reason the implications of resource use efficiency on policy and market efficiency/efficacy, and vice versa, are never far away (Leibenstein, 1966, both distinguished and linked the two, plus see Rogers et al., 2002). Daly (1992) refers to resource efficiency in economic-allocative and fiscal terms; in other words how society successfully bears down on an environmental or allocation problem by selecting the 'proper' tool (e.g. pricing), applying this at an appropriate scale and cost and observing outcomes. This is summed up by Tarlock and Wouters (2007): 'The transcendental objective of efficiency requires that the resource be allocated to the most valuable suite of uses'. Young (2012, p.10) draws the same conclusions but uses three different terms for 'efficiencies' quoted here:

- Productive efficiency because water can be deployed to the place where it makes its greatest contribution to the economy, society and the environment.
- Allocative efficiency because investment takes full account of long-term risks, opportunities and expected changes in supply and demand.
- Administrative efficiency because transaction costs are low and recovered from users.

The paracommons framework regards the first as 'economic efficiency' and includes a fourth type of efficiency 'resource use efficiency' (the latter in turn incorporating elements of resource productivity). UNEP (2012) define four types of interrelated efficiency concepts: technical efficiency, productive efficiency, product-choice efficiency and allocative efficiency. The additional 'product-choice efficiency' is, according to UNEP, a local scale reflection of allocative efficiency whereby resource users draw from a number of price and technological information signals to make investment choices to address their technical efficiency.

While it is expedient to exclude Daly's and Young's three governance dimensions common to all questions of resource sustainability and instead focus on physical efficiency within resource management, they are all interlinked in ways that add to rather than resolve the complexities suggested by REC and the paracommons. These four together offer a comprehensive interpretation of 'efficiency as governance' – the subject of an entirely different yet worthy book on the various dimensions of efficiency (Singleton, 1999). For example, Poulton *et al.* (2006) draw the mutually beneficial links between allocative/ economic efficiency and resource efficiency. However, as I argue in Section 4.4.2, the uncertainties of resource efficiency management also stem from mixed terminologies and definitions. This might explain why in their essay, although arguing for clarification, Olschewski and Klein (2011) appear, to my mind, to mix analyses of resource efficiency and economic efficiency, and why Wildavsky, in 1966, cautioned for a teasing out of the different constructions and purposes of efficiency.[13]

3.3 Efficiency and spatial scale

This section employs two related topics to draw attention to efficiency risks arising when spatial scale is introduced (where I use Gibson *et al.*, 2000, to distinguish between different types of scales). The first topic examines single technologies at one level of production – for example, efficiency within the household or industrial unit. This can be contrasted with the second sub-section, addressing various scale-related issues.

3.3.1 Efficiency as single technologies in single units of production

Efficiency draws the attention of disparate parties (resource users, officers, engineers, scientists, chief scientists and so on) and yet a considerable amount of interest is levied at high-profile successes that exist as single technologies for single units of use and activity. A good example is when the household is viewed as being the seat of innovation and transformation via the adoption of low-energy light bulbs without seeing energy consumption arising from collections of households or from other patterns of, say, commuting, consumerism, and consumption. Single technology examples can also be found via the UK's Environment Agency 'Waterwise' awards for water efficiency including single garden watering packs, eco-showers, water-reducing hand basins, rainwater harvesting and metering.[14] It is also possible to interrogate high-profile claims by international drinks companies that they are conserving water when closer inspection reveals their competence is in reducing water consumption of bottling plants rather than the embedded water of the sugarcane that goes into drinks (the latter comprising about 90–95 per cent of the water in the production chain). While these technologies might 'save' some water for the unit, few ideas are assessed, by peer review, for their reduced basin consumption of water, or how they might be scaled up, or how they reduce consumption at the unit level in the long run, or how consumption of linked resources such as energy are not increased.

3.3.2 Efficiency as scale-related or sector-related technologies

Following on from the topic of single-unit successes in raising efficiency, there is also the question of an appropriate scale of interest for examining and communicating efficiency-and-recycling thinking. The question of scale has many dimensions explained in the book but two points can be made – first is that terminologies and methods within one sector (for example, water) often apply to one level but don't transport well to higher levels. For example, at the crop-level, I believe, for example, the species-dictated metabolic efficiency of converting water into crop biomass (giving rise to the term 'water use efficiency') while of interest to irrigation scheduling within systems (Payero et al., 2009) has less relevance for managing irrigation systems (using classical irrigation efficiency) or for determining the impacts of irrigation systems on the hydrology of catchments (where effective irrigation efficiency is the better concept). Likewise the argument that small units of irrigation (say 1 to 10 hectares) fed by drip irrigation should be adopted to provide an appropriate solution to basin-wide irrigation could drive up energy and maintenance demands.[15]

Second, at the larger scale, one must be cautious about over-extending the boundaries of the systems to be included. For example, I surmise that we may never completely trace and quantify the complexities of an extended water commons incorporating all the different sectors and uses of water, and here I refer to the full gamut of 'flows' of water in and out of wetlands, rainfed agriculture, soil water, groundwater, hydropower, water and sanitation, and wastewater irrigation from human settlements. Although these systems will have elements of reuse and waste/wastage, to lump them together to create a 'meta' (or mega) picture of resource use efficiency may lose more than is gained. Each of these types of water use will have their own versions of resource efficiency complexity.

Similarly, when assessing combinations of resources such as energy and water, analyses will have to proceed carefully if combined numerators of benefits, and combined denominators of consumed resources are to retain informational utility for decision-making. Although analytical tools incorporating exergy and emergy (Odum, 1996) help convert different resources to fewer metrics for efficiency and productivity type analyses, these remain thermodynamically (kilojoule) based – a conversion step that might involve loss of information regarding how the resource itself is managed. In the case of irrigation in the paper by Chen *et al.* (2011), unexplained assumptions regarding how water was saved enabled their research to conclude it was 'water savings' from efficiencies of 45 per cent increasing to 70 per cent afterwards that resulted in improvements from project interventions. It is my belief that their paper's failure to explain how these efficiencies were determined that undermines the validity of their conclusions – and signals the risks created by simplified efficiency assumptions applied to resources that have their own detailed 'efficiency theory' discourse (CAWMA, 2007; Perry, 2007; Lankford, 2012b; van Halsema and Vincent, 2012; Pereira *et al.*, 2012).

3.4 Efficiency and socio-ecological systems

A brief look at efficiency within other types of systems offers ideas that the paracommons draws upon and yet provides a useful contrast to the paracommons. In the interests of space, three types of systems are examined: industrial ecology, ecosystem services, and the water-energy-food nexus.

3.4.1 Efficiency as industrial ecology and recycling

One field that has developed a view of the practice of efficiency and management of losses is that of industrial ecology (Bourg and Erkman, 2003; Ayres and Ayres, 1996; Socolow *et al.*, 1996; Boons and Howard-Grenville, 2009). Drawing on ecology and ecological studies, various definitions of industrial ecology are available – chiefly by minimising impacts on environmental resources and recycling resources between organisations. Although parallels exist between the paracommons concept and industrial ecology thinking, there are six substantive differences outlined and explained in Figure 18.

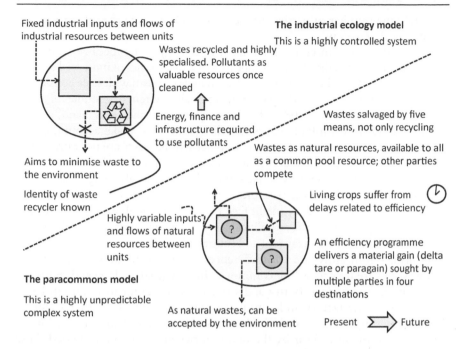

Fixed industrial inputs and flows of
industrial resources between units

The industrial ecology model

This is a highly controlled system

Wastes recycled and highly
specialised. Pollutants as
valuable resources once
cleaned

Energy, finance and
infrastructure required
to use pollutants

Wastes salvaged by five
means, not only recycling

Wastes as natural resources, available to all
as a common pool resource; other parties
compete

Aims to minimise waste to
the environment

Identity of waste
recycler known

Living crops suffer from
delays related to efficiency

Highly variable inputs
and flows of natural
resources between
units

An efficiency programme
delivers a material gain (delta
tare or paragain) sought by
multiple parties in four
destinations

The paracommons model

This is a highly unpredictable
complex system

As natural wastes, can be
accepted by the environment

Present ⟹ Future

Figure 18 Comparing industrial ecology and the paracommons

First, efficiency in industrial ecology is achieved by the practice of recycling. However, in the literature I have surveyed, the effects of efficiency gains over time (in dimensional and dimensionless forms) are less-well explored, theorised and assessed. As identified in Section 4.5.2, the paracommons understands that there are five main ways efficiency may be arrived at (forestalling, recovering, avoiding, offsetting and transferring losses) whereas industrial ecology is predominantly interested in recovery via recycling. Thus the paracommons is predicated on efficiency gains for systems under transition, delivered by policies designed to raise performance, which in turn generate material gains that are 'up for grabs' by common pool competitors. Material gain generated by an efficiency gain is not an explicit part of industrial ecology.

Second, the remits of the paracommons and industrial ecology are natural and industrial resources respectively. In this case, I take 'natural resources' as those mediated by ecosystem services (e.g. water, soil, air and nutrients). I understand 'industrial resources' as being a combination of mined mineral resources (such as iron ore, sand, coal) and manufactured resources (such as steel, pig iron, silica, electricity). Natural resources convert to living entities (worms, crops, trees) and industrial resources (iron ore, silica) convert to manufactured goods (car batteries, smartphones). This resource distinction introduces both the greater toxicity of the wastes involved in industrial conversions and the specialised nature of recycling and re-manufacture associated with industrial ecology. These constrain the number of parties that seek out and compete over the wastes – a point picked up below.

Third, the paracommons and industrial ecology differ because scale and nestedness are treated differently. It appears to me that studies of industrial ecology mostly examine flows between parts of a recycling organism with scale considered only when recycling is a scaled-up version of a smaller system (e.g. studies of global waste flows, Hotta *et al.*, 2008). The paracommons, on the other hand, obtains its complexity from multiple scales nested together that make accounting and management at one scale problematic. In other words, losses are not only picked up by a neighbouring system at the same scale/level but by a wider encapsulating system at a higher scale. However, like symbiosis in industrial ecology (Paquin and Howard-Grenville, 2009) the paracommons invokes the idea of neighbourliness where one system's efficiency might feed or impair a neighbouring system. Also with scaling up comes a raft of political economy issues affecting whole sectors under transition between a current and future efficiency.

Fourth, significantly, are the differences arising from how losses, wastes and wastages are treated as 'the commons'. In industrial ecology and industrial metabolism, studies (e.g. Bourg and Erkman, 2003) point to pollution being the export of one factory and simultaneously the input of another factory – however, in this process, considerable industrial resources (finance, machinery, energy) are brought to bear by the two factories in order to make good those pollutants by a number of means – for example, filtration, re-smelting, cleaning and de-oxidising to name a few. Because of the nuisance value of the original waste pollution and the cost and technological barriers to enter the common pool competition, the wastes of one factory are not sought by all users – and particularly not by the environment unless water treatment is expressly funded. On the other hand, in the paracommons, losses usually comprise harmless or relatively harmless materials (e.g. water vapour and drainage water) that are valuable and sought by all users if salvaged. In addition, the paragain might be claimed by others not associated with or neighbouring the efficiency-raising effort. It is this common interest that differs from the specialised interests of industrial recycling. Though as technologies for waste become cheaper or more adaptable to non-factory situations, or the waste materials become financially more valuable, then common pool competition for the resources held in pollution is set to increase – boosting similarities with the paracommons.

Fifth, certainty and predictability is different. In industrial ecology, inputs into the system are controlled by the nature of the business environment, future contracts and chains of production. The flows of these resources into and between industrial units are known, quantified and pre-set. In the paracommons, inputs are stochastic and naturally variable. Being 'natural' resources, dynamic and exogenous drivers create conditions where systems are in non-equilibrium and where efficiency is highly relative to those conditions. In the paracommons, 'saved' resources are generally not known because of the problems of measuring the current and future systems that also flux and change greatly.

Sixth, timing of resource flows differ between industrial ecology and the paracommons. Exemplified by irrigation, because crops are living entities growing in limited soil-water the timing of delivery is important and is influenced by efficiency. This timing element is not precisely reproduced in industrial ecology which instead views timing as an economic rather than life or livelihood-bearing problem.

3.4.2 Efficiency as industrialising ecosystem services

I draw a distinction between the industrial focus of the previous section (industrial ecology) and a focus on industrialising ecology or ecosystem services (MEA, 2005). I am particularly interested in how science and society interprets ecologies for different objectives and purposes over time. Table 9 shows a sample of ecological purposes cast as provisioning, regulating and cultural ecosystem services. Five 'forest' ecosystem services are selected: biodiversity; biomass accumulation for regulating river flow regimes; oil palm production; wood pulp and timber production and carbon sequestration.

Table 9 Interpretations of various forest ecosystem services

Forest purpose	Numerator 'service'	Denominator 'losses'	Managerial focus or tool to achieve this
Biodiversity	Cultural services as traditional landscapes and amenity value: species diversity, also helping with regulating services, next.	Timber not extracted but could be is a 'loss'.	National park enclosures.
Biomass accumulation for river flow and regime	Regulating service: pristine rainfall-runoff response – high base flows. Achieved by building up forest biomass and permeable forest soils. Leaf litter and recycling.	Timber not extracted is a loss.	National park enclosures.
Oil palm	Provisioning service: palm oil.	Biodiversity is diminished.	Private forest plantation.
Wood pulp and timber production	Provisioning service: harvested timber for paper.	Timber wastes, leaf litter, biodiversity are losses.	Private forest plantation.
Carbon sequestration	Regulating service: timber for long-term wood products. Biochar.	Short-term timber products are losses. Soil fertility for agro-ecology is not maintained.	Mixed use and management.

Thus a forest previously protected for its ability to sustain forest villagers, pristine river flows and biodiversity might be 'switched' to a production system with a focus on wood pulp or palm oil. If we accept an efficiency/productivity equation for each of these services, this means the numerator of the efficiency equation changes. In the example just given, the ecosystem service numerator on top of the ratio switches from being biodiversity to wood tonnage. This change in emphasis then has consequences for what divides between the denominator (the input side) and the numerator (the outputs) and what becomes the 'losses' (services that are less significant). To recap, with a changing view on which ecosystem service is to be boosted (thereby industrialising ecology) comes a parallel change for which ecosystem service or products are viewed or deemed as losses (depending on one's perspective). Defining inputs and outputs become central to the question of the efficiency ratio and how it may be adjusted because both numerators and denominators are in turn shaped and influenced by a complex conversion of supporting ecosystem services and forest inputs and processes around light, water, soil nutrients, microbial enzymes, atmospheric gases.

The significance of switches in ecosystem service and associated losses for the paracommons, paragains and liminality is picked up in following chapters. These switches set up a spectre of complexity in terms of efficiency and productivity – throwing light on the question of who gets the gain and who loses during these changes in emphasis.

3.4.3 Efficiency mediated by resource connections

One of the 20 complexity factors explored in the next chapter (labelled 'resource nexus') involves connections between energy, water and food. Tangible and important correlations between irrigation efficiency and and the consumption of other resources are illustrated by Karimi *et al.* (2012a) in their analysis of the energy usage in accessing and distributing groundwater in Iran. Their argument is that reducing the consumption of water saves energy usage and contributes towards lowering carbon emissions. While further work on this topic is required, this argument supports the idea that classical irrigation efficiency has utility beyond its engineering design function and should be considered in a wider societal context.

By connecting irrigation efficiency to other resources, and to an ever-changing context of water supply and demand, one can caution against policy that seeks to optimise the efficiency of one resource (e.g. water via supporting drip irrigation) while neglecting other scarce resources such as power and energy or farmers' skills and disposition towards managing and maintaining complex infrastructure collectively.

Furthermore with a wider cross-resource view it is possible to parenthesise the value an absolutist 'hydrologist' viewpoint on water accounting which van Halsema and Vincent (2012) usefully draw attention to. Foster and Perry's

(2010) call for proper hydrological accounting is partially useful in so much as other livelihood, system properties, economic and resource implications are excluded from this. However by including these parameters, irrigation efficiency has to be approached with specific reference to the particular circumstances of the irrigators, system and basin being studied. Having modelled a variety of contextual factors, Medellín-Azuara *et al.* (2012) also make the same point in their conclusion.

3.5 Efficiency and perspective

The paracommons is interested in different scales of production from the field level up to the basin or beyond to society as a whole. This necessarily incorporates a variety of 'stakeholders', each having particular perspectives. To address this issue, three subsections examine: a socialised approach to efficiency, disciplinary differences, and briefly a political ecology of efficiency. This topic is discussed futher in Chapter 7.

3.5.1 Efficiency as social partnership

The influence of support services working alongside irrigators, particularly irrigation engineers, in irrigated agriculture is considerable. This topic is beyond the remit of this book (see Zwarteveen, 2010) but nonetheless a few points may be made. It is my interpretation that despite much movement towards a 'socialised model' of irrigation (characterised by Wageningen University) over the last two decades, involving participatory irrigation and irrigation management transfer, large swathes of irrigation in developing countries fall under the responsibility of irrigation and affiliated professions (government and non-government engineers, think-tank researchers and university academics). The result is that programmes to boost irrigation efficiency are dominated by engineer normative thinking and formal training (Chambers, 1988) that irrigation efficiency should be raised by engineering solutions such as canal lining, metering, and sprinkler and micro-drip systems. Whether efficiency might be seen through the eyes of farmers and their lives and livelihoods still seems to be largely missing. Furthermore, technological ingenuity, making previously 'lost' resources salvageable, depends on a considerable interplay of knowledge, creativity and skills between different resource stakeholders. Engineers do not hold the monopoly on solutions to water efficiency.

One facet of the debate on irrigation efficiency is a propensity to misjudge efficiency levels achieved in mathematical percentage terms (notwithstanding considerable data gaps and uncertainties). Political ecologists might recognise the actor-orientation of this propensity – views on efficiency and efficiency numbers depend on their provenance: donor, regulator; expert engineer, upstream farmer; downstream farmer; and so on. This accommodation is embodied in the research design and work by Lopez-Gunn *et al.* (2012) and van

Halsema and Vincent (2012) who clearly point to the social study of irrigation. Furthermore Knox *et al.* (2012) distinguish multiple views on the basis of 'tests' and introduces a disciplinary argument; that when applied by farmers, efficiency takes on an economic rather than purely hydrological character.

As these authors show (supported by Boelens and Vos, 2012 and Lankford, 2012), putting farmers at the centre of the efficiency debate may have profound consequences for the choice of the model and modelling of irrigation efficiency improvements. Lankford (2006, 2012b) argues that judgements on irrigation efficiency should capture local excellence in water management as a baseline, rather than the unachievable target of funnelling 100 per cent of abstracted water through beneficial crop growth. Similarly, van Halsema and Vincent (2012) argue that a focus on relative differences in water productivity in a given context helps to switch low productivity practices to high productivity practices, rather than focusing on absolute high values. Medellín-Azuara *et al.* (2012) draw attention to the trade-offs farmers might make between capital expenditure and recurrent costs tied in with crop and farming system types. This exercise opens up researchable questions about the nature, size and cost of the technological step-change required to reduce water consumption.

I add one coda to this section should readers interpret the paracommons to be predominantly a systems-type framing that downplays people and communities. It is my sincere belief that a future understanding of (irrigation) efficiency will distinguish resource user knowledge on water management (including their definitions of misuse and abuse) from formal professionalised (e.g. engineering) perspectives. There can be no more sobering way of seeing efficiency than through the eyes and experiences of those working with resources and understanding what frames and mediates their approaches. But this requires critical interdisciplinary field research to distinguish between what individual irrigators might parochially understand as priorities for them (which may not transform performance) from difficult-to-discern system improvements that might benefit farmers and their irrigation indirectly. It is this 'teasing out' of system improvements with the full support of farmers and engineers that provides the challenge in order to benefit collective systems at field, tertiary, secondary, scheme and basin levels. This critical participatory model is an intensely political choice about the democracy of resource governance (Woodhouse and Nieusma, 2001).

3.5.2 *Efficiency as sector-specific; translating paradoxes*

Moreover the treatment of the ramifications of efficiency in the literature is highly sector- or discipline-specific, with one consequence being that a comprehensive framework has not been agreed despite efforts by scholars (Jollands, 2006). Outside the debate on irrigation explored in this book, perhaps the best known is the Jevons Paradox commonly found in the energy literature (Herring, 2006; Sorrell, 2009). Yet although the word 'paradox' echoes the uncertainties of the paracommons, the Jevons Paradox is discussed in terms of how more energy gained from an energy potential translates into 'efficiency-induced consumption of outputs' (Polimeni *et al.*, 2008) by creating cheaper and more energy. Here, the paradox arises because with improved throughput efficiency comes lower costs combined with an improved likelihood to consume – in other words efficiency savings leads to a 'rebound effect' of greater total consumption (also Sinn, 2012). However, as I make clear later, irrigation efficiency drives uncertainties in different ways to that of energy because irrigation has many material pathways that conversions can follow leading to changes in both locations and proportions of distributions of resources and salvaged losses.

While resource recycling addresses a multitude of consequences of efficiency improvements between the 'upstream' wasting sector and 'downstream' reusing sector, the nature of 'paradox' should be circumscribed . In wastewater irrigation (Qadir *et al.*, 2010; Wichelns and Drechsel, 2011) or urban waste recovery (Gandy, 1994; Gutberlet, 2008), this relationship (though explored in many ways) is relatively static rather than highly dynamic and uncertain. In this literature, the waste product is a common pool resource flowing from an upstream 'creator'. While complex, there is not much 'paradox' in this one-way relationship.[16] Indeed following on, one must not believe paradoxes lie everywhere. While the paracommons concept is predicated on the ingredients for paradox being present (the space between current assumptions, future expectations and to future outcomes), these will not apply in all circumstances.

3.5.3 Resource use efficiency as political ecology

I take political ecology as the study of the framing of narratives of environmental governance and management through the co-production of knowledge, theory, policy and practice (Blaikie and Brookfield, 1987; Rocheleau, 2008). Political ecology studies how 'environmental truths' become aligned with protagonists exploring authority, organisational provenance, and the interests and education of key individuals. Resource efficiency can be viewed in similar way; as a contested debate in environmental science with competing ideas and complex genealogy:

- In irrigation, contest is most acutely seen in the debate around the nature of local losses, and whether, by being 'captured' by the basin, they should be taken into account in an efficiency type calculus or simply volumetrically portrayed as a fraction (not in ratio). This in turn feeds debates about the nature of water accounting (as seen in recent issues of JWI).
- A number of papers in the *Agricultural Water Management* 2012 Special Issue on efficiency discuss a variety of concerns over definitions of efficiency and associated terms such as 'waste' and 'losses'. This reflects the consternation shown by other authors (Perry, 2007; Seckler *et al.*, 2003) over lax deployment of terms. Thus authors Pereira *et al.* (2012), observing

poor rigour in the use of definitions of some key concepts such as water use efficiency and water productivity, formulated a framework of terms supported and specified by equations, and usefully distinguish between losses and waste.
- The relationships between definitions, historical development, hydrological computations and political ecology are aptly expressed in van Halsema and Vincent's (2012, p. 13) conclusion:

> Since the 1940s, the interests of study in irrigation efficiency have moved from studying water application, through irrigation system design and operation into system performance evaluation and basin water accounting. In these shifts, variables which once had a specific contextual meaning have become related to the economic and social context of the study of water use. They can also be related with how groups of people make claims and steer policies to justify or transform water use for users and the environment. It is this political use of factors, that are derived and also often portrayed as neutral, that makes them relevant to study under the frame of political ecology.

Although a political ecology analysis of efficiency is not the overt purpose of this book, the paracommons concept allows efficiency views that are compartmentalised or over-idealised to be interrogated. Politics and political ecology are picked up again in Chapter 7.

3.6 Chapter review

The objective of this chapter was to reveal the diversity of views that exist within the seemingly misunderstood subject of natural resource efficiency. Whether a consensus on approaches to a theory of, and policy towards, resource use efficiency is possible in the future will depend on many considerations, one of which, in my view, is an examination of factors that bestow complexity – the subject of the next chapter. While other authors might distinguish more characterisations, efficiency can be seen in different ways:

- Although in recent months the topic of efficiency appears to be growing in significance, efficiency has not been as fashionable as other natural resource indications such as justice and equity, adequacy, sufficiency and access.
- Physical efficiency is an inherent part of biological and economic resource productivity and vice versa – recursive connections exist between dimensionless and dimensional ratios of inputs to outputs. Three fields of efficiency were briefly explored: economic, biological and physical/water management.
- Efficiency is a functional part of systems design allowing gross design dimensions of abstraction and conversion infrastructure to be computed.

- Characterisations of modern and traditional designs imbue systems with over-simplified notions of performance.
- Efficiency is employed as a performance measure for single units/ components as well as a performance measure of whole, larger and encompassing systems. Taking this point, efficiency is scale-related. Calls to improve efficiency and recycling might mistakenly see efficiency as residing within individual units without viewing the whole system where paradoxical knock-on effects might reside.
- Adopting an industrial ecology lens, emphasis is on recycling of waste and wastages within industrial complexes to minimise end-of-pipe externalities.
- Efficiency may be viewed as discipline-related and as a social partnership; views on efficiency depends on one's provenance and relationship with the resource.
- Efficiency is drawn into sharp focus by resource connections; between land, water, energy and labour to name a few.
- Resource efficiency plays a significant underlying part of the regulatory, administrative and technical governance of natural resources, aiming to improve economic efficiency and drive allocation of resources towards more profitable and productive uses.
- The material gains generated by raising efficiency of natural resources conversion can lead to distributive concerns about 'who gets the gain of an efficiency gain'. Thus alongside supply and demand management, efficiency takes up a role in 'share management'.
- Illuminated by political ecology theory, efficiency is interpretable through the above points yet shaping the nature of the relationships between people, agencies, policies, resources and systems. Efficiency can be employed by governments as a modernising 'project' rationalising technological interventions to change modes of production.

Chapter 4

A framework of resource efficiency complexity

4.1 Introduction

The more the use of a resource is defined by the presence of losses and efficiencies characterised by quantity, economic value, geographic scale(s) over which efficiencies take place, and by a multiplicity of perspectives, inputs, conversions and pathway options, the greater the complexity of resource use efficiency. With this complexity comes a greater likelihood of incomplete understanding, inadequate monitoring, erroneous accounting of resource flows and ill-designed interventions designed to reduce consumption or change distributions via efficiency and productivity. And with these risks come broader differences between the hopes, potentialities and eventualities of policy interventions – hallmarks of the paracommons. In this chapter I explain how resource efficiency complexity (REC) arises and gives rise to the paracommons marked by properties of liminality and inter-user connectivity.

In terms of the formulation of this approach to resource management, I have hypothesised components or 'building-blocks' of complexity (Schneider and Somers, 2006) organised within a framework. The framework given in Figure 9, and detailed in Figure 19, fits this description; '[The framework] is inclusive of various systems, processes, and scales. The job of the framework is to provide a roster of components, and to suggest how the components may relate to one another' (Cadenasso et al., 2006, p. 4). Figures 9 and 19 show the framework schematically. I would expect the framework proposed on these pages to be familiar to the epistemic community working on complexity and social-ecological systems. Rammel et al., (2007, p.12) express the terrain of the conceptualisation of complex adaptive systems:

> Natural resource management systems as complex adaptive systems (CAS) are characterised by their dynamic interdependencies across various scales and are driven by mutual interactions between institutional, ecological, technological and socio-economic domains. Hence, we argue that sustainable management requires interdisciplinary analysis and improved understanding of multi-dimensional feedbacks and, more generally,

of the dynamics of the interrelations between the particular interacting subsystems.

4.2 Environmental, economic and political context

The environmental, economic and political context shapes and is shaped by four dimensions of resource use efficiency. First, mentioned in the outer ring of Figure 9, a number of contextual factors increase the significance of efficiency in the search for economic sustainability and environmental protection (Keys *et al.*, 2012). Many of the following environmental changes have solutions in, or are partly addressed by greater efficiency and productivity: increasing resource scarcity; the capping of resource consumption to protect ecological systems; allocation of resources; the search for green growth and environmental sustainability; the increasing value of, and ability to recycle, wastes and wastages; growing recognition of the linkages between resources such as land, energy and water, and; the location of efficiency in responding to natural variability and dynamics such as drought and flooding.

Second, the changing environmental and economic context not only sets interests in efficiency (the previous point), it also establishes and affects existing efficiency performances and the manner in which local resource users respond to new stimuli. Thus context also mediates the extent to which losses are materialised and controlled. An example is where forests shift in emphasis from being utilised for local communities for a variety of forest products to a system that focussed on long-term carbon sequestration. This shift delivers subtle changes in the nature of lesser important forest products or, in other words, wastes and wastages.

Third, context influences the objective of efficiency savings; what the material gains from efficiency reworking are subsequently utilised for – and thus to whom they are delivered. As discussed in Section 4.5.3, four destinations have been identified. These are: to boost the efficiency and productivity of a proprietor system; to boost the productivity of the immediate neighbours of the proprietor; to conserve the common pool of resources, and; to improve the distribution and performance of resources within a wider economy.

Fourth, rapidly changing physical, climate and political contexts feed into the uncertainties surrounding the interpretation of efficiency as discussed in Section 4.5.5. By being in, and contributing towards, a dynamic environment of contraction and expansion (e.g. dryness and wetness, economic boom and recession) efficiency is rendered more complex and difficult to judge. But countering this, it is because of environmental and economic variability that we gain experience in judging efficiency measures that accommodate change rather assuming absolute measures are static and have wide validity.

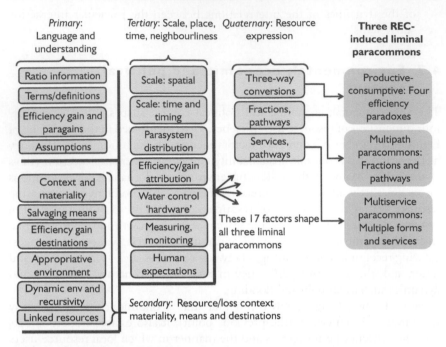

Figure 19 Twenty sources of resource efficiency complexity in four groups

4.3 Sources of resource efficiency complexity

In examining how the resource efficiency complexity and paracommons are created, I identify 20 'building blocks' or sources of complexity. The 20 sources are grouped in an arrangement of four types, shown in Figure 19, classified simply as primary, secondary, tertiary and quaternary.[1] Primary sources mostly deal with issues arising from the definitions, language and understanding of efficiency. Secondary sources principally refer to the nature, context and purposes of the resource efficiency and savings. Tertiary complexities arise when accommodating scale, nestedness, neighbourliness and time. Quaternary origins address resource expression – three are proposed: first, for resources that mainly resolve into three outcomes (exemplified by burning coal giving useful heat; ash waste and heat/gas wastage). Second, for resources undergoing efficiency conversion processes that can fall to very different pathways but which have only one or two forms (a good example is water and vapour on irrigation systems). Third are resources that produce diverse ecosystem services (exemplified by forests as carbon and forests as biodiversity).

It should be re-iterated that while the primary, secondary and tertiary sources influence all three types of liminal paracommons and underpin a wide variety of terminology, knowledge, social, spatial and scalar issues, the quaternary class of resource expression give rise to three types of liminal paracommons. Because the three types of paracommons are also described later in Chapter 5 the reader is also referred to text and diagrams found further ahead in the book.

Attention is given to an explanation of the paracommons and REC arising through system dynamics and behaviours rather than to expend effort exploring in great detail the social, political and power complexities related to managing the commons and commons transitions (Berge and Van Laerhoven, 2011; Geels, 2010). Nevertheless, the social, political and scientific contexts and understandings of these issues are also sources of complexity. In other words one might argue each of the 20 sources can be interpreted in turn via a political ecology and sociological understanding of efficiency. Aside from being potentially tautologous, these perspectives might best be reserved for future reflections. Nevertheless, Sections 4.6.7 and 7.1 of this book address some of the social and political dimensions of efficiency.

4.4 Primary: definitions, language and understanding of efficiency

4.4.1 Ratio information loss

Ratio information loss is the first of 20 sources of complexity. Information loss describes the frustrating lack of information imparted by an efficiency ratio, leading to uncertainty regarding what is in ratio to what, and whether the purpose of the efficiency ratio is understood by all (see also Chapter 4 in Princen, 2005). This means that what appears to be a purely technical and straightforward ratio ends up being a matter of interpretation (selecting the benefits and losses to go into the numerator and denominator). This is in addition to the interpretation required to judge efficiency results (in other words what does, say, '67 per cent efficient' mean?). Thus resource use efficiency begins with the appearance of what appears to be a simple ratio between what is withdrawn from natural capital and what deemed beneficial for society – the latter comprising benefits in the shape of goods, processes or services. An example might be (staying with water), a river as natural capital, an irrigation intake withdrawing some water and the crop transpiration as the beneficial 'process'. However, both the amount and type of benefit and losses that define the calculation of the ratio are negotiable and subject to other priorities. For example, in deficit irrigation the amount of water beneficially transpired in the numerator is purposively and sometimes subjectively reduced.

Thus the efficiency ratio, as explained below, hides a considerable array of underlying natural processes and stages that in turn should obligate natural resource scientists to clarify what inputs and outputs are being assessed. An efficiency ratio is appealingly simple – yet creates risks for conveying information about multifaceted, variable and diverse social-ecological systems. It is this potential loss of information about the system under study that requires scientists, managers and users to be vigilant about the number and construction of efficiency ratios when assessing and communicating changes in resource production and performance.

At the heart of efficiency complexity, the ratio of the conversion of a natural resource to desirable benefits contains a vinculum between numerator and denominator that both binds, but obfuscates and blinds the relationships between inputs and outputs. The vinculum is the name of the line '/' in the ratio 'a/(a+b)'. Importantly, resources pass over the line or threshold from the denominator in the common pool to the numerator as a produced good or service. First and foremost, confusingly for the unwary manager, efficiency can increase from changes to both numerator and denominator. In addition, the inclusion of part of the numerator within the denominator (as with irrigation efficiency) leads to fuzziness about what parts or stages of the conversion are in ratio to each other.

Efficiency ratios can trap the unwary, swaying professionals towards making summary judgements rather than seeing efficiency as a complex process entailing a material loss and gain of resources. As a 'performance measure', the efficiency ratio conveys the effectiveness of a conversion process; thus an 85 per cent efficient system is more effective at converting resources to benefits than a system with an efficiency of 75 per cent. However, one might forget the efficiency ratio also conveys information about total quantity of losses (wastes and wastages) within a system; in other words 85 per cent efficiency signals 15 units of loss per 100 units withdrawn, and what this subsequently means for salvaging those losses. This matters when we start to compare systems. To explain, an irrigation efficiency of 75 per cent for a small irrigation system of 120 hectares comprises a smaller volume of water losses than an irrigation efficiency of 85 per cent for a system of 3000 hectares. This distinction underpins discussions in later sections on the distributive objectives of efficiency and on offsetting losses from one system of production to another in order to raise aggregate (e.g. river basin) efficiency.

Furthermore, multiple stages, materials and pathways within a conversion process might be obscured by a single ratio. To illustrate, Figure 37 later on in the book, depicts a flow from left to right, moving from natural capital to the production of a good. Central to the efficiency ratio is the conversion of inputs to outputs generating some waste or wastage. It would be possible to generate one overall ratio of the benefit on the right-hand side to the input on the left-hand side. Yet Figure 37 offers two intermediate stages; withdrawal and consumption. At each stage, unrecoverable wastes/wastages (light grey boxes) and recoverable wastes/wastages (black arrows and boxes) are produced. If re-drawn specifically for irrigation (rather than generically as it is now) Figure 37 would have some boxes and arrows added and removed to depict further details. For example, withdrawals into an irrigation system account for beneficial consumption by crops (transpiration) plus all recoverable and non-recoverable fractions, plus non-beneficial consumption (non-crop evaporation). Consumption adds together beneficial and non-beneficial consumption before leaving only beneficial consumption remaining. Furthermore, irrigation losses can become an input to another irrigation system. What appears to be a linear sequence of losses ends up being circular and recursive.

Retaining clear information on the individual parts of the sequential, parallel, circular (and lifetime) process of producing a good or service is not easy. Resource management may be seen as a series of parallel practical activities and resources coordinated to appropriate, grow, harvest, store, process and therefore convert various natural capitals such as land and water into products. Examples of activities include scheduling water, selecting varieties, adding fertiliser, ploughing soil and so on. These inputs are subject to their own conversion processes which if biological (rather than, say, chemical or industrial), turn nutrients, water, sunlight and atmospheric gases into living things. Derived from these inputs and conversions are intermediate outputs such as rice grown, timber, wild plants and fish. And from these, following further conversions of storing, refining and processing, are final products such as hulled rice, packaged wild products and frozen fish ready for sale.

Therefore analysing and comparing performances of resource management means defining inputs, outputs and ratios, recognising these are fraught with implicit assumptions about the distillation of complex and multiple processes into simple indices – such as tonnes, area, cubic metres of water, labour-days, dollars and so on. These indices are utilised to create technical performance ratios of outputs to inputs of two types: as dimensionless 'efficiency' ratios or percentages (e.g. the irrigation efficiency of smallholder irrigation system) or as dimensional efficiency or 'productivity' (e.g. tonnes rice produced per hectare). Easily forgotten are animate/inanimate related timing aspects of efficiency if the conversion process depends on timing to keep things alive and growing – for example, crops on irrigation schemes (Lankford, 2006). Much of these sector or system specific aspects of efficiency are often lost in the choice and distillation of information to a single 'efficiency' figure.

In addition, ambiguities are often increased if efficiency type ratios are taken for granted. In industry and commerce, 'eco-efficiency' views manufactured products or services as the numerator, with one or more 'environmental pressures' as the denominator WBCSD (2000, p.9), making eco-efficiency a measure of resource productivity, rather than a dimensionless efficiency indicator. This helps to explain the attraction of eco-efficiency in that the denominator is aggregate impact on natural capital, but it introduces processual and informational ambiguities if the numerator is not carefully considered or that the *means* to raise efficiency and productivity and subsequent distribution of salvaged resources are not unpacked. This is a criticism I level at irrigation productivity (Lankford, 2006, 2012) – that while crop tonnage per water volume usefully integrates many steps within the denominator it also results in a loss of information about water management within an irrigation system. This means one might have the paradoxical situation where high yields from fertilisers, tillage and seeds mask poor water management.

Given that this REC framework also addresses changes in emphases of ecosystem services it is worthwhile introducing this topic in this section on 'ratios'. In defining which service takes precedence as the 'benefit', we (society and science) consequently select which services have secondary value and therefore are deemed to be 'losses'. An 'agro-forest' that previously prioritised nuts, fruit and timber (as its numerator) sees a change in emphasis when the same forest is switched to permanently sequester carbon which, in the form of long-lived timber or biochar, becomes the numerator. The corollary of this change in the numerator is that the denominator (containing the losses) also changes. In the agro-forestry case, losses are sub-standard nuts and fruits (or that eaten by other species) and poor quality timber, but in the latter 'carbon' case, even good quality timber might be deemed to be a 'loss' if it is not sequestered. These changes highlight the political nature of efficiency ratios as a result of changing emphases. This topic is particularly discussed in Sections 4.7.3 and 5.2.4.

4.4.2 Hazards of ambiguous terms and definitions

A minefield of efficiency terms and definitions creates difficulties in understanding exactly what is being discussed and analysed. A lack of definitional precision incorporating user perspective (what is gross demand from an irrigation system equates to net withdrawals from the river), scale/level (e.g. differences between household efficiency and wider efficiency in the economy apply, see Sorrell and Dimitropoulos, 2008) and subject- or disciplinary use of the English language offers multiple, competing definitions in this subject area. In addition, terminology confusions might arise because subjects are at different stages in their development of thinking and associated definitions.

Until agreement over definitions is achieved (not the purpose of this book), talking at cross purposes will be a signature of negotiations over resource use efficiency and the paracommons (see also Neuman, 1998, for the role that vague definitions play in negotiating new water rights on the basis of water conservation). Terms such as 'loss', 'saving', 'waste' are used throughout in the irrigation and other literatures ill-advisedly, carelessly by rote or wilfully for gain. Or one might rightly defend the use of terms on the basis of spatial, scalar or system innocence; for example, irrigation managers may understandably seek to quantify the efficiency of their 'local' system but be less interested to know how their systems impact on the wider river environment.

Drawing on definition problems in irrigation efficiency and productivity, and for the purpose of introducing the liminal paracommons theory, I am cautious in using both lay and precise terms. Critical readers will know that, for example, 'saving a water loss' is useful in a vernacular, introductory and perhaps local or even policy sense but limited in a precise accounting and scientific sense because 'lost water' may be recaptured downstream. Furthermore, for the purpose of discussing the distributive effects of raising resource use efficiency, it is important to distinguish between 'efficiency' per se, (a dimensionless ratio of performance) and the material losses and gains from a conversion-and-loss process – a point made above. As discussed in the next few sections, materiality matters; it provides a focus on the question of who gets the 'saved losses' from an efficiency improvement programme. Another problem is that 'loss' and 'gain' are words that often go together to represent a trade-off rather than a material translation of one into another.

Reflecting on these points, and by referring to Table 10, I suggest that 'losses, wastes and wastages' is an introductory or descriptive phrase where losses equal wastes and wastages combined. However, because the words 'wastes' and 'wastages' are associated with pollution, the less pejorative terms 'losses', 'tare' and 'extrinsic fraction' can also be employed. 'Tare' is a general term and 'extrinsic fraction' is the more scientific term. Both terms are identical and both cover 'losses' and 'wastes and wastages' combined. Adding the losses to the intrinsic fraction for the production of the good/service/process arrives at the gross or total amount withdrawn. 'Losses' and 'tare' also work better than 'wastes and wastages' when ecosystem services are switched in emphasis. In other words, when timber harvesting from a forest is promoted over pristine biodiversity management, then timber becomes the numerator 'service' and biodiversity becomes one of the denominator 'losses'.

Tare may be explained thus: if an irrigation system is 65 per cent efficient (in the classical sense), then gross withdrawal is 100 per cent, the net consumption (or intrinsic withdrawal) for beneficial goods, services and processes is 65 per cent, and the tare or extrinsic part is 35 per cent (which may be further divided into recoverable or non-recoverable losses or other fractions). Thus, I distinguish intrinsic withdrawal of a resource from a common pool from extrinsic withdrawal. Tare is equivalent to the extrinsic portion and is there to meet expected losses so that the intrinsic requirements may be met wholly.

One other point on the intrinsic fraction must be addressed. The intrinsic volume of natural capital is not fixed and objective. Not every system requires a uniform or fixed demand of resources for the production of standardised goods (see previous text on retrenchment, subsitution and dematerialisation). Thus in irrigation there is a difference between full irrigation (providing crop water requirements completely thereby maximising returns to land) and deficit irrigation that stresses crops often leading to minor yield decreases but yet 'saves' the water applied to the field. The point is that 'savings' can be made by reducing both the intrinsic and extrinsic fractions. The subject of deficit irrigation and its impact on savings and timings of water delivery is so central to the complications of the paracommons that it is picked up again in Section 4.6.4.

In addition there are possible linguistic confusions over the words recoverable and recovered.[2] One question is whether both are volumetrically the same and whether 'recoverable' means conditional and 'pending' or is an observed fact after a given activity. Although it adds to the mental agility required, I favour not making them the same because separation cautions against a mass-balance logic that a recoverable flow is always recovered.

Table 10 Navigating terms associated with losses, wastes and wastages

Term	Term	Lay terms	Lay terms	Fractions terminology	Recovered or recoverable	Consumed or not	Examples in irrigation
Gross amount or Withdrawal	Intrinsic fraction	Net	Goods, services, processes	Beneficial consumed fraction	Not recoverable by definition	These two fractions are consumed	Crop transpiration
				Non-beneficial consumption	Not recoverable by definition		Soil evaporation
	Extrinsic fraction	Losses or tare	Wastage				
			Waste	Non-consumed fraction	Recovered	Wastes are not consumed but may be recovered	Reused drainage water
					Not recovered		Water to sinks not then used

Table 10 recognises that not all stakeholders, scientists and readers are at the same stage of understanding or see systems from a collective high-level unit such as the river basin. The lay terms on the left-hand side of Table 10 may not satisfy those who have moved to fractions (Perry, 2007) on the right-hand side as a way of dealing with accounting problems associated with the words 'losses', 'wastes' and 'wastages'. The difficulty is in creating a set of satisfying interchangeable lay and precise terms.[3]

4.4.3 Paragains: material gains from changes to efficiency

In this section I introduce material gains and their conceptual location. A fuller calculation of paragains is contained in later sections. Continuing with the theme in the previous section on definitions, and in researching efficiency, I am particularly struck by the limitations of the English language (and science vocabulary) in discussing two related aspects of 'losses'. There appears to be a lack of terms to describe the differences between the *losses* of a conversion process and how, when these losses are reduced, this might lead to *gains* of a resource (in other words, when 'saving' creates something released or 'freed up'). A problem of comprehension occurs when we become interested increasing a system's efficiency to 'free up' (or 'save') a loss/waste/wastage to create a resource that can be put to further use. As above, terms such as 'free up' and 'save', useful in everyday parlance, are limited in a scientific sense. In this section I explain how improvements to efficiency and productivity performance that salvage losses generate 'new' material gains (as resources) and then explain some issues that cloud and confuse that logic. Because this idea of a 'new' resource is so central to the paracommons, the reader is referred again to Figure 7 as well as Figure 20.

The term I use to cover a material gain is 'paragain'. The term 'paragain' denotes a very high level of calculative, definitional and managerial uncertainty. In short, a paragain cannot be easily verified, 'freed up', tracked and donated to a particular party. The terms 'salvaged losses' and 'material gain' are nearly equivalent but the reason salvaged losses are not identical to material gains (and therefore paragain) is because what might be salvageable might not be salvaged, and furthermore, a paragain also covers the resources freed up when the net demand (in fractions terms: 'the beneficial consumption') is also reduced when substitution, dematerialisation or retrenchment takes place. However, for the remaining text of this section, 'salvaged losses' are discussed as an approximation of paragains.

The logic of finding a 'new' resource from an efficiency performance gain requires a system to undergo a time-shift. Thus, the left-hand side of Figure 20 shows the current (and supposed) 'inefficient' system and the right-hand side shows the hoped-for more efficient system in the future. The loss or 'tare' in the current inefficient system is greater than a smaller loss/tare in a future more efficient system. This gives a difference in the amount of losses when moving

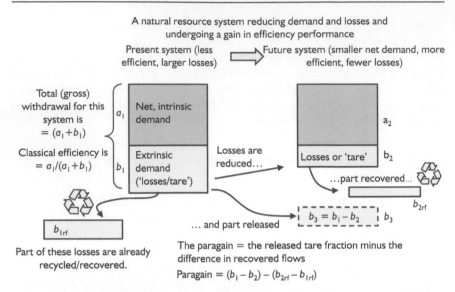

Figure 20 shows a diagram with the following labels and text:

A natural resource system reducing demand and losses and undergoing a gain in efficiency performance

Present system (less efficient, larger losses) → Future system (smaller net demand, more efficient, fewer losses)

Total (gross) withdrawal for this system is = $(a_1 + b_1)$

Classical efficiency is = $a_1/(a_1 + b_1)$

a_1 Net, intrinsic demand

b_1 Extrinsic demand ('losses/tare')

Losses are reduced...

a_2

Losses or 'tare' b_2

...part recovered...

b_{2rf}

b_{1rf}

... and part released

$b_3 = b_1 - b_2$ b_3

Part of these losses are already recycled/recovered.

The paragain = the released tare fraction minus the difference in recovered flows

Paragain = $(b_1 - b_2) - (b_{2rf} - b_{1rf})$

Many other calculations for paragains exist depending on how resources are saved, their effect on aggregate depletion, and where they end up. This is why the calculation of resource paragains is complex and debatable.

Figure 20 Efficiency performance gain leading to a material gain

from 'today' to the future, which can be depicted as 'delta losses' (or 'delta tare'). This delta tare can potentially be a resource that can then be claimed by parties. In other words the material gain comes from the difference between current larger losses and future smaller losses. A greater step-change from a very inefficient current system to a very efficient future system would mathematically free up larger losses (a larger 'delta tare'). In addition to the losses 'freed up' one can also via substitution, dematerialisation or retrenchment, decrease the amount of the net material consumed. Although this is not shown in Figure 20, a lower consumption of a resource per unit of 'good produced' is why resource productivity might increase under such circumstances (as happens with deficit irrigation).

However, this logic is coloured or confused by a number of other factors simultaneously taking place. For example, the time-shift 'delta gain' has to be discerned against the current-day recycling of losses to other parties – particularly back to the common pool (and as shown later on, also to a downstream neighbour). The recycling of losses is shown in Figure 20 towards the bottom of the left-hand side – using the circular 'chasing arrows' recycling icon. This distinction introduces not only confusions for the unwary but also a scientific challenge over the extent to which efficiency gains are feasible if losses are already being recycled to other users. The debate in the recent Water International papers (Frederiksen and Allen, 2011; Gleick *et al.*, 2011;

Frederiksen *et al.*, 2012) is partly about this vexing question. Gleick *et al.,* (2011) believe gains are achievable while Frederiksen *et al.,* (2012), it appears, argue that gains are either minimally available because losses are already being recovered or because any gains are immediately recaptured by the proprietor irrigator to extend his or her irrigation. The corollary of this latter behaviour, however, demonstrates that without an extendable farm area to irrigate, the gain would pass to a neighbour - be it irrigator or other downstream user.

Scientists therefore face various potential confusions and incongruities. One confusion sees losses, wastes and wastages as something polluting and negative (and therefore to be reduced) but – and this is the incongruity – as something that is productively beneficial (a positive) that can be 'increased' in an efficiency-raising programme. Confusion also arises from distinguishing between losses already recycled 'in the here and now' from losses that can be recovered and found anew by a shift in management performance. Another complication arises because recycling can reduce in response to changes in productivity of the loss-generating system. While forestalled and recycled losses result in resources being 'salvaged' there are significant differences in the management effort for recycling and the effort for raising performance (Section 4.5.2 explains the five ways losses are salvaged). In the recycling case, one can 'simply' allow outcomes at the basin level to capture the extent to which recycling has reduced consumption and raised efficiency (hence the shift from classical to effective efficiency). In the case of forestalling, reducing and releasing losses, much more careful water management case is required at the farmer and field level. However, how to achieve this is by no means agreed and certain – particularly when scaling up improvement programmes to many thousands of farmers.

Because this resource gain from a resource loss is the defining characteristic of the paracommons, more explanation is now given. To accommodate and explain these points I use a variety of terms to describe how a loss fraction of a resource conversion-and-loss becomes an efficiency-related recovered material gain. Figure 21 helps this discussion in seven stages. To begin with the left-hand side of Figure 21 (column A) depicts the common pool from which a resource will be withdrawn. This might be a dam, aquifer or river. A part of the common pool (CP) remains throughout the seven stages. However, in column A the withdrawn fraction has a dotted line. As made clear elsewhere in this book, a water right for an irrigation system already has within it an accommodation for water losses. This expectation of losses (often not carefully measured or designed for) is an extremely important part of the paracommons.

Column B1 in Figure 21 adopts column A, but differentiates more clearly between the withdrawn resource and the remaining common pool resource. Also in column B1, sitting within the withdrawn resource are fractions (RF, NRF, BC and NBC; Perry, 2007; see Section 4.7.2) that comprise the dispositions of the withdrawn amount.

We then move to Columns C and D which dispense with fractions instead employing wastes and wastages (explanation is given below). Using dictionary

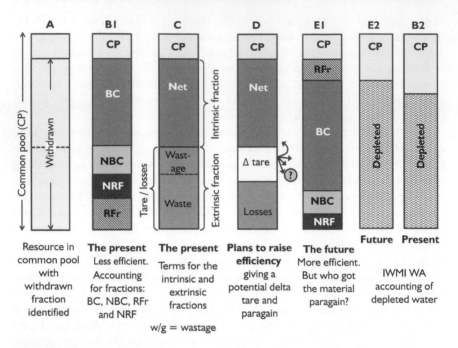

Figure 21 Following efficiency losses and gains from the present to the future

definitions, wastes are recoverable materials and wastages are non-recoverable materials (noise, vibration, vapours/gases) with little or no economic or productive use. Wastes are equivalent to non-recovered and recovered fraction while wastages translate into the non-beneficial consumption. Also in Column C, in addition to the terms 'wastes', 'wastages' and 'losses', the less pejorative terms 'tare' and 'extrinsic fraction' are employed. Thus Column C follows the classification given in Table 10 above. Column C also represents the current system prior to any efficiency programme.

Column D contains the policy intention to raise irrigation efficiency and 'save' losses. This column shows sees the basin from the proprietor's point of view. In other words, plans to (implicitly) alter the balance of 'common pool water versus total depletion' are expressed through programmes to improve the efficiency of irrigation systems and reduce wastage rather than the other way around. (The latter would involve selecting the proportion of depleted water to water remaining in the common pool and then pursuing efficiency programmes to meet those plans.)

The next column, E1, is the outcome of this policy if things go well and according to plan now analysed in terms of resulting fractions. Column E1 represents the achievement of a) higher efficiency and b) reduced loss fractions and c) resource preserved in the common pool (CP plus recovered flow). Column E2 then shows how these fractions transpose into the IWMI fractions of preserved/utilisable in the common pool and depleted. E2 can then be compared with B2 (B2 is the basin-level depletion pattern of Column B1 prior to any efficiency programme being implemented. Comparing E2 and B2 gives a reasonably accurate picture of any material gains expressed at the basin level, and in this case the gain in B2 has moved to the common pool.

But we need to return to Column D which sits between the present (C) and the future (E1) and contains the white box termed 'delta tare'. Given the complexities of managing efficiency in large socio-ecological systems, and the likelihood of an under- or mal-designed efficiency programme, this material gain, in terms of its volume, timing, location, ownership and quality, cannot be view with any certainty. On the contrary, it should be seen as the potential gain arising from a system in transition; or put another way, as the paragain. Thus the white box in Column D, 'delta tare', must be understood at all times for it to transpire according to the intentions of any efficiency raising plan. Accordingly, it needs to be accounted for, both quantitatively and in terms of loss of quality or delayed timing. Then it must be accounted for in terms of final destinations, for example, either to the proprietor system or back to the common pool or to other users (hence the four small arrows and question mark). However, as explained later in Section 4.5.4 under 'appropriative environment' the distribution of this paragain often goes to the proprietor system.

It is now possible to explain how the stages in Figure 21 contribute to efficiency complexity. When re-working systems, the intrinsic and extrinsic fractions do not necessarily keep separate and neither do different extrinsic fractions; they can switch into each other. For example, wastage (evaporation from a pool of water sitting on open bare soil in a cropped field) could, by adding or maintaining drainage, be turned into a recoverable waste (drainage flow). Furthermore, a newly salvaged resource (the delta tare in Column D) may distribute to different owners – hence the question mark in Column D – and this may change the current routing of recovered losses. For example, it is possible that previously recovered waste under an improvement programme is, in the future, beneficially consumed with the result that less of the resource ends up with recyclers or the common pool. This means that changes in tare are not always 'towards the ecologically positive' but can also result in a reduction in downstream flows. (Note this would not happen if an increase in beneficial consumption was drawn instead from non-recoverable or non-beneficial losses.)

Using Figure 21, we can now see where water accounting plays a role (introduced in Chapter 3, and discussed in more detail below in Section 4.7.2) and how the paracommons framework complements water accounting. The Perry water accounting methods are applied to the 'current' and 'future' stages in Columns B1 and E1 respectively, while E2 and B2 show the final two outcomes expressed at the basin scale as seen through the IWMI WA methodology (Karimi et al., 2012b), returned to the common pool or fully depleted. In terms of explaining the paracommons alongside water accounting:

- Although as currently framed water accounting might be able to discern current and future patterns of final dispositions of water withdrawn, it does not lend itself to interventions that attempt to raise efficiency and subsequently guide where salvaged resources might be sent to.
- Fractions do not discern how loss fractions are coupled together – that it is difficult to adjust one alone without affecting other fractions. For example, reducing excess watering of a field reduces both runoff at the tail end of the field (likely to be a recoverable flow) and standing water evaporation which is non-beneficial consumption. Water accounting cannot discern the detailed ways that water might flow through an irrigation system. The paracommons, as currently expressed, also struggles with this problem, but it is interested in the finer details and dynamics of water management.
- Fractions are not common language amongst water users and managers looking to improve the efficiency and performance of their particular system. The paracommons attempts to grapple with terms and ideas that have different meanings for different actors.
- The idea of a potential 'liminal' paragain is not found in water accounting. It arises in the paracommons because efficiency-related programmes are by their nature interested in a future where performance is (supposedly) raised and water saved. This is the normative aspirational starting point that many policy-makers and engineers (and sometimes farmers) have. But there are likely to be differences between the original aim to raise efficiency and final outcomes – the question is how and what accounting systems get closer to understanding and predicting likely outcomes.
- A continuation of the theme of Figure 21 is covered in Sections 4.7.2 and 5.2.3 where it is revealed that there are many different ways that a paragain can be calculated depending on the boundaries set for the system.

Furthermore, there are vexed questions of the ownership (Neuman, 1998) over the tare part of the resource not yet wasted, as well as the resource wasted, plus the resource subsequently 'saved' if an efficiency programme is implemented and veritably creates 'real' material gains. Thus reuse and the reduction and translation of the delta tare from one user to another modifies the principle of subtractability applied to natural resources (that resources subtracted in one place are not available elsewhere) requiring revised thinking on common pool resources. These points are picked up again elsewhere in this book.

4.4.4 Prevailing assumptions about losses, wastes and wastages

Connected to the expectation of a resource gain discussed in the previous section, are outmoded yet 'normal' (widely accepted) assumptions about loss, waste and wastages. To explain this, I first wish to convey the sense that for resource users within a dominating political economy, interests in wastes or by-products are changing, practically, theoretically and politically, from a position of little interest to one where interest and value is increasing. This change results in resources having a *current* loss/waste/wastage fraction destined to be 'wasted' and claimed by nobody yet *in the near future* could be valuable and the subject of intense interest.

As indicated elsewhere, this changing interest in 'losses' is driven by a set of interlinked changing circumstances such as scarcity, technological improvements, their consumption by another, and markets and prices. But this change from prevailing assumptions to new purposive interests is not a smooth and uniform process. A tension arises between current custom towards what *is deemed* to be uninteresting, waste/wastage, and a dawning realisation of new waste/wastage values in a fast-moving and increasingly scarce world. This tension is perhaps one of most subtle, even psychological, aspects of the paracommons. In addition, the process of moving with the times, so to speak, will not be happening to all resource stakeholders simultaneously. For example, some irrigators quickly adopt more efficient technologies while others do nothing remaining with existing practices influenced by systemic constraints or the assuredness of water rights and possibilities for further rent-seeking (Dellapenna, 2002). Furthermore how wastes and wastages are measured and accounted will be central to claim and counter-claim.

It is the contrast between assumptions regarding efficiency and extant outcomes that generate paradoxes associated with the field of resource efficiency. In addition to resource users' assumptions, simplified and/or anachronistic views are also held by scientists and policy-makers. These assumptions, as Crase and O'Keefe (2009) argue, shape poorly-informed policies designed to drive resource users to adopt new practices. For example, in irrigation, it is assumed that 60 per cent of water goes to 'waste', whereas that fraction may be much less. Thus, irrigation is treated for forthcoming losses which do not exist and may not reduce withdrawal – leading to ineffective spending of development aid (for example, refer to Lopez-Gunn *et al.* (2012) and Crase *et al.* (2009), or to government policies behind drip irrigation subsidies in Syria and India, to name two.

There is a more subtle dimension regarding assumptions surrounding losses, wastes and wastages. Assumptions around the identity of 'losses' described so far are driven by shifting understandings in turn driven by scarcity, competition or by refinements in terminology. This ignores a deeper ontological question about the very nature of 'loss' in a conversion of a natural resource to a good or service. Subjectivity in defining wastefulness may not be resolved easily – and grades between what comprises fully beneficial conversions (such as water to produce 15 tonnes of sugar per hectare from sugarcane) to conversions that are part beneficial (water consumed by trees that line and stabilise an irrigation canal). This subjectivity must also be understood in terms of systems and cultures of production; to compare canal systems with closed glasshouse hydroponic systems simultaneously misses the point (hydroponics cannot be scaled up limitlessly or 'speak' to the global scale of irrigation management) but also makes the point about the good or service being pursued alongside 'the loss'. High-humidity, high-temperature glasshouse tomato production means that the ratio of water held in the final product compared to that beneficially transpired through the growth of tomatoes is higher compared to open field production. In other words, in the glasshouse system the 'good' is the water inside the tomato whereas in the field, it is the soil water transpired. This shines a light on the common understanding of 'fractions' that underplays the amount of water taken away in the final product.

In the question of ecosystem services and losses, there is an even more complicated discussion about the identity of losses if all services are seen as co-beneficial and co-existing. For example, what is 'a loss' or 'inefficient' in a forest simultaneously producing food and timber, sequestering carbon and mediating river flows? One ecological answer is nothing (ecology/biology does not waste; Phillips, 2008) yet managing ecosystems services towards specific societal interests tells another story (Leach *et al.*, 2012). For the purposes of illustrating the paracommons, I can only set up relatively clear examples of the switches and emphases afforded to different ecosystem services.

Assumptions to date on this question are largely guided by long-standing and slowly evolving professional and disciplinary norms and debates. In theory 'fractions' should allow a penetrating discussion about the very 'soul' and nature of a conversion and loss – but not as currently (and contentiously) being conducted in water journals where the 'fractions' protagonists seem not to wish to understand another party's views. While the paracommons framework is a contribution to the debate, I too am not using the paracommons to directly analyse the identity and ontology of different resource conversions-and-losses (see Princen, 2005, for more).

There are also epistemological questions regarding efficiency assumptions. I return to the problem of how to derive a common set of linguistics and assumptions that navigate the benefits of a *normative* idea of efficiency (that resource managers should seek to be more efficient in managing complex systems), the *positive* idea of efficiency (that resource managers need to know what is going on and how their efficiency experiments change resources and users for better or for worse) and a *critical* idea of efficiency (that complexity, errors and competing views shape results and policy requiring greater study than hitherto conducted). Navigating this would allow some resolution of the sustainability paradox of achieving efficiency without increasing resource consumption. Furthermore all these ideas need to be approached with greater criticality, empiricism, scepticism and imagination, asking how we set aside a focus on somewhat meaningless percentage efficiency figures (at least for irrigation) but deliver much more meaningful water management. How do we address 'green paradox' concerns regarding efficiency-induced higher consumption via a more intense study and governance of how material gains flow to four different destinations (Section 4.5.3)?

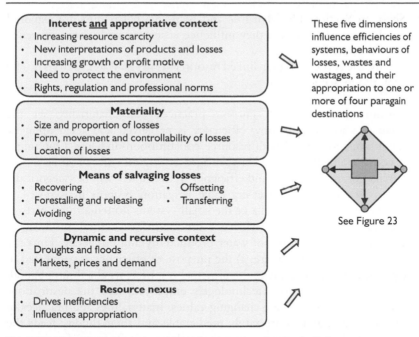

Figure 22 The context and materiality of resource savings and efficiency

4.5 Secondary: context, means and paragain destinations

In this section, I discuss how the dynamics of efficiency, losses and paragains play out in a given context. The section is particularly interested in how context, materiality and the means of salvaging losses drive up an interest in losses and then subsequently guide any resource paragains to different 'destinations'. Seven efficiency dimensions are relevant to this section (Figure 22).

- First, context defines the extent to which losses become the focus of heightened interest. This topic was addressed in Section 4.2, and in the outer ring of Figure 9 and is not returned to in this section.
- Second, the materiality of losses influences their 'salvageability' and to some extent to where salvaged regained resources flow. This topic is covered in Section 4.5.1.
- Third, the five means of salvaging losses and creating 'paragains' are discussed also with reference to how they guide the appropriation of paragains by different parties. This is the subject of Section 4.5.2.
- Fourth, the four paragain destinations are discussed in Section 4.5.3.
- Fifth, context is examined for the manner in which it is appropriative: how and why paragains are biased towards one or more of the destinations. See Section 4.5.4.

- Sixth, in Section 4.5.5 dynamic and recursive behaviours are explored for the manner in which they influence efficiency and where efficiency paragains terminate.
- Finally, the context of inter-linked resources (energy, water, land) is briefly explored in Section 4.5.6.

In summary, the context, current practices and materiality set the circumstances and terms of how efficiency is perceived and acted upon by the many players associated with resource consumption-and-loss. One might observe considerable differences between agro-ecological types or zones in this regard. Irrigation management and efficiency in the Andes will be categorically different to that faced by Soviet-era canal irrigation systems in Kyrgyzstan, ex-colonial canal systems of India or the small systems dotted around parts of savannah sub-Saharan Africa. Each has a unique infrastructural, professional and social ecology and culture of water productivity and efficiency. Thus the three messages of this section are; a) the purpose of efficiency is connected to the reciprocities between context, the principal resource, its chief output/good/ service, its waste/wastage/loss, technologies employed, and the destinations for the salvaged losses; b) these changing values, materialities and practices of losses, wastes and wastages are rarely predictable and their fluidity feeds the uncertainties of the paracommons and; c) paradoxes arise because the observed destinations of paragains do not correspond with our expectations of where they should go.

4.5.1 The materiality of losses, wastes and wastages

The materiality of the losses matters as this has a bearing on how society seeks or is able to control, salvage and relocate them. Table 11 discerns three main factors: tangibility, quantity and location. All things being equal, a greater amount of salvaged losses will come from: tangible wastes rather than intangible wastages; greater amounts of waste and wastage; and losses located in places where they can easily be recovered and/or forestalled. The earlier discussion on definitions hints at the ambiguities between wastes (physical or chemical material often seen as polluting though recoverable) and wastages (often intangibles such as gas, heat, noise, vibration). However, these might interchange as wastes and wastages become less or more valuable which in turn feeds through to technological and market ingenuity. Thus in the face of scarcity and value, a previously worthless wastage (not recovered by anyone) switches to a recovered waste.[4]

The materiality of ecosystem services and of the degree to which the 'loss' (newly moved to the denominator by a switch in emphasis) continue to be co-benefits would depend on particular circumstances. Carbon dioxide is instructive because it is ostensibly a wastage product from an industrialised society yet vested with value via carbon markets. Being fixable via photosynthesis in forests for offsetting purposes, CO_2 becomes a common pool for conversion

Table 11 The materiality of losses, wastes and wastages

Factor/Property	Interest and ability to control losses, wastes and wastages	Significance for paragain destination (Section 4.5.3)
Form/tangibility/ controllability	Influencing the extent to which wastes or wastages are 'made physical' so that humans or nature can make use of them by control and redistribution. Affects timing, flows, qualities of wastes and wastages.	Affects distribution and allocation because the wastes and wastages can be controlled, divided, moved and directed.
Waste/wastage size and proportion	The size of the loss, waste, wastage in comparison to total resource availability.	Influences all four destinations/purposes.
Location of resources and of wastes and wastages	Influences where wastes/wastages start and end up. For example, water losses from irrigation can enter aquifers.	Affects distribution and allocation.

into goods (such as timber) and recoverable and recovered 'wastes' (such as smaller branches, bark and leaves). This turns upside down the nature of the 'commons' since now atmospheric carbon dioxide becomes the 'commons' and carbon in sequestered long term in forest timber or soil biochar becomes the beneficial product following conversion. See Sections 5.2.4 and 7.3 for more on this discussion about forests and atmospheric carbon.

4.5.2 The five means of adjusting efficiency and salvaging a loss

This subsection addresses the five means of adjusting efficiency and salvaging a loss. I employ the word 'salvage' to cover all five ways and for simplicity's sake, the term 'salvaging a loss' is approximately the same as 'generating a paragain'. (There are extra dimensions to paragains that come from particular attention paid to substitution, dematerialisation and retrenchment – and these are explained where appropriate.)

In broad terms, efficiency should be understood as a process involving the conversion of resources to a good or service alongside the production of wastes and wastages. However, given the number of resources (water, energy, land, labour, chemicals, etc.) and the myriad ways in which 'savings' may be made within and across individual conversion processes, it would be beyond the remit of this book to itemise the unique procedures and steps by which efficiency can be adjusted. Instead, recognising the conceptual aims of this framework, I theorise five main means by which losses can be salvaged with the consequence that aggregate efficiency is raised. The five means are forestalled/ released, recovered/recycled, avoided, offset and transferred. Details are given in

Table 12 The five means of adjusting efficiency and salvaging paragains

Means	Explanation
Forestall and release losses	Losses, wastes and wastages are reduced within the process of conversion and loss leading to a released 'delta tare'. This might happen throughout the growing season or be focussed within one part of the growing season in order to compensate for losses at another time.
Recover losses	Wastes are recovered and reused. In water, this would normally take place by gravity flow of water from one system to another system. Unrecoverable wastages could be converted to wastes and subsequently recovered.
Avoid and release losses	Wastes and wastages are reduced and released by being avoided altogether. This entails a reduction in withdrawal of the principal resource from the common pool in order to reduce the net intrinsic consumption via substitution, dematerialisation and retrenchment. Thus a reduction in production is effected to reduce coupled losses that are difficult to forestall or recover through other methods.
Offset losses	Wastes and wastages are addressed elsewhere in a related neighbouring system with the consequence that total losses within the 'parasystem' are salvaged. This means that the efficiency effort is transferred to another system. Thus the original proprietor system remains running unchanged.
Transfer production and losses	A situation which arises when parts or whole of a production system are moved to another location with the consequence that coupled and other associated losses are moved as well. In this instance, production and losses are avoided at the original proprietor system.

Table 12, with worked examples given in Table 13. Tables 20 to 22 (pp.148, 150, 154) give examples of how losses can be forestalled and released.

The 'offset' and 'transferred' options are examples of 'efficiency pathways' based on water neutrality and offsetting (Hoekstra, 2008) where extrinsic losses are not altered in the system under study but are changed in an associated system within the same river basin. In carbon sequestration, distant offsetting via carbon credits is a much more important part of the productivity problem of managing the carbon paracommons. The offsetting of efficiency and 'wastes' between very different and geographically separated systems hints at the complications of delivering efficiency gains over large areas. Offsetting both water and carbon require heightened levels of verification to test the claims of additionality (that these exercises would not have been adopted regardless, and that they genuinely reduce water depletion and boost sequestration). While water allows mass-balance equations such as those in Table 13 to be calculated, computations for carbon-based ecosystem services are much more complex because they are complimentary and not divisible and additive.

The starting position in Table 13 is of two neighbouring systems (X and Y) each with their own withdrawal from a river, and each having a classical irrigation

Table 13 Examples of five means of raising efficiency and salvaging losses

| | Current scenario | | | Future scenarios: reworking two irrigation systems; attempting to be more efficient and/or less consumptive | | | | | | | | | | | | | | |
| | Starting point | | | Forestalled to X: NBC, RF and NRF reduced | | | Recovered to CP: RF up, NBC and NRF down | | | Avoided: X withdraws less; CIE stays same | | | Offset: X efficiency same, Y efficiency up | | | Transferred withdrawal; X less; Y takes more | | |
System	X	Y	Agg	X	Y	Agg	X	Y	Agg	X	Y	Agg	X	Y	Agg	X	Y	Agg
WD	100	100	200	100	100	200	100	100	200	50	100	150	100	100	200	50	150	200
BC	45	45	90	65	45	110	65	45	110	22.5	45	67.5	45	75	120	22.5	110	132.5
NBC	20	20	40	10	20	30	5	20	25	10	15	25	20	5	25	10	20	30
NRF	15	15	30	10	15	25	5	15	20	7.5	5	12.5	15	5	20	7.5	5	12.5
RF	20	20	40	15	20	35	25	20	45	10	35	45	20	15	35	10	15	25
Check total	100	100	200	100	100	200	100	100	200	50	100	150	100	100	200	50	150	200
Tare	55	55	110	35	55	90	35	55	90	27.5	55	82.5	55	25	80	27.5	40	67.5
Consumed	65	65	130	75	65	140	70	65	135	32.5	60	92.5	65	80	145	32.5	130	162.5
Depleted	80	80	160	85	80	165	75	80	155	40	65	105	80	85	165	40	135	175
CIE	45.0%	45.0%	45.0%	65.0%	45.0%	55.0%	65.0%	45.0%	55.0%	45.0%	45.0%	45.0%	45.0%	75.0%	60.0%	45.0%	73.3%	66.3%
EIE	69.2%	69.2%	69.2%	86.7%	69.2%	78.6%	92.9%	69.2%	81.5%	69.2%	75.0%	73.0%	69.2%	93.8%	82.8%	69.2%	84.6%	81.5%
Delta BC	Paragain to X, Y, Agg			20	0	20	20	0	20	-22.5	0	-22.5	0	30	30	-22.5	65	42.5
Delta depletion	Paragain to the CP/WE					-5			5			55			-5			-15

Key: WD = Withdrawn. BC = Beneficial Consumption. NBC = Non-beneficial consumption. NRF = non-recovered fraction. RF = recovered fraction. Tare = NBC+NRF+RF. Agg = aggregate from both systems X and Y; CIE = classical irrigation efficiency (CIE = System BC/System WD); EIE = effective irrigation efficiency (EIE = System BC/System total consumed); Ag CIE = Aggregate classical irrigation efficiency (Ag CIE = (Aggregate BC/Aggregate WD). Ag EIE = Aggregate effective irrigation efficiency (Ag EIE = Aggregate BC/Aggregate consumed). Note: for illustrative purposes and lack of space, other possible fractions are not shown – though these are tested later in the book.

efficiency of 45 per cent applied to 100 units withdrawn giving each 55 units of 'loss/tare'. Of these 55 units, 20 in each system are recovered to the river basin and 35 units are depleted via non-recovered flows and non-beneficial consumption, giving an aggregate total depletion over the two systems of 160 units. No paragain is calculated because no 'improvement' thus far has been applied.

In the forestalled scenario, the 55 units of loss in System X are reduced to 35 units, giving a classical irrigation efficiency of 65 per cent. The NBC, NRF and RF all fall to 10, 10 and 15 units respectively. This means that the paragain of 20 units went to System X and of this, 15 units came from NBC and NRF and 5 units were 'taken' from the common pool that previously was returned to it. The aggregate parasystem classical and effective efficiency is 55 per cent and 78.6 per cent.

In the recovered/reused scenario, System X increases to the same higher efficiency of 65 per cent but this time changes its proportion of NBC to NRF. Thus 35 units of total loss in this future scenario are split into 5 units of NBC, 5 units of non-recoverable losses and 25 units of recoverable losses. In this scenario, the proprietor gets another 20 units and an extra 5 recoverable units result in a paragain of 5 units to the common pool.

In the avoided scenario, System X withdraws 50 units instead of 100 units (through retrenchment). Thus 45 per cent efficiency applied to 50 units means 27.5 units are the loss (tare). This means that the proprietor loses 22.5 units but the common pool gains 55 units (=160 depleted in the current scenario minus 105 depleted in the avoided scenario).

To examine the offset scenario, we first see that in the starting scenario, both X and Y withdraw 100 units and both have an efficiency of 45 per cent. In X, efficiency savings are problematic and costly. Savings are offset to the neighbouring system, Y to compensate for X. To free up gains from both systems, requires system Y to raise efficiency up to 75 per cent. The outcome is that System X proprietor sees no change while System Y gains 30 units, while the common pool sees a reduction of 5 units.

In the transferred case, similar to an offset, the more capable system Y now withdraws 150 units transferred from the less capable System X which now withdraws only 50 units. In terms of the distribution of paragains, System X drops 22.5 units and System Y gains 65 units and the common pool also sees a reduction of 15 units.

Therefore, Table 13 hints at different aspects of complexity and uncertainty in trying to re-work systems: a) the assumptions and expectations that drive attempts to move from a current to a future efficiency; b) the need to refer to combined effects of two or more systems if offsetting and translating of efficiency and production is taking place; c) the various trade-offs between the 'tare fractions' within the five means to salvage resources resulting in different efficiencies of the two systems X and Y, as well as the combined system; d) uncertain final dispositions and destinations of paragains including distribution of water to farmers within systems X and Y.

How the five means tend to be employed to steer resource paragains to one or more of the four destinations is discussed in the next section.

4.5.3 Four efficiency paragain destinations

Picking up on the previous section and linking to forthcoming discussions on scale and neighbourliness, an important factor introducing scientific and political uncertainty, paradox and space for manoeuvre is the identification of the destinations of the 'freed up' or saved amounts of resource (paragains). By inquiring about an efficiency destination, we are simultaneously asking about the purpose of salvaging a resource and who or what gets (or owns) the 'saved' fraction. Put simply, if the paragain destination is the common pool, then the purpose of being more efficient is a reduced consumption of natural resources. If 'purpose' is not asked and answered clearly enough, efficiency policy might not meet expectations. One of the reasons that efficiency is seen as paradoxical in not reducing the consumption of natural resources is because we predetermine the saved fraction will pass to the common pool of resources. It is this contrast between our ecological rationale for efficiency savings and counter-intuitive observation of increased consumption that provides the paradox (Crase and O'Keefe, 2009; Ward and Pulido-Velázquez, 2008).

The 'whom or what' suggests an purpose or owner-destination for the waste or wastage 'saved'. Table 14 and Figure 22 introduce the four main destinations of a paragain freed up by an improvement in efficiency. In other words, the salvaged paragain moves to or is held within one or more of the four destinations/parties. These are: 1) for use by the proprietor or single system to boost productivity and performance; 2) for users immediately neighbouring the proprietor system for their productive benefit; 3) for the common pool and environmental conservation and productivity, and 4) to boost consumption and production in the wider economy and society. All four destinations are grouped together in a quadrangle termed a parasystem. Table 15 also explains which of the five means of making savings applies most to each of the four destinations. By considering scale and neighbourliness, further discussion of the destinations and parasystem is picked up in Sections 4.6.1 to 4.6.3.

Prior to explaining these four objectives or destinations, it might be worth exploring another interpretation of a destination for savings – that of 'banking' a gain. This is where a saving is put into a storage body such as a dam or aquifer for later reuse. In some senses, banking a gain might also be seen as paying back a debt or loan of water which in turn provides further options for how savings may be managed. However, for the time being, 'banking' is incorporated with each of the four destinations. A dam therefore might belong to a proprietor system in which case the saved water is 'owned' already and is highly likely to be reused within the proprietor system. A dam or aquifer might also receive saved water then subsequently used as an environmental good or as water for economic and societal purposes. Also in

Table 14 The four paragain destinations and means of 'salvaging' resource losses

Sector or system destination	Explanation of objective/destination of the salvaged fraction	Means of salvaging losses
Resource gain for the production of the proprietor or single system	Emphasis on process efficiency, production and performance from a single systems point of view. The proprietor system is the one already withdrawing the resource and 'causing' the losses. Implicitly or explicitly the freed-up fraction is incorporated into higher production of goods and services within the proprietor system.	Forestalled/released Recovered Offset Transferred
Neighbours get the paragain boosting equity and productivity	With reference to resource users sharing a particular neighbourhood (e.g. tertiary canal system), the gain is taken by users or by dominant users simultaneously which then leads to a different shared dynamic between users; more equitable and more productive. Immediate downstream neighbours are included in this destination.	Forestalled for parallel or neighbouring users Recovered quickly by downstream neighbours
Common pool or environmental conservation / productivity	A forestalled or recovered fraction is returned to, or kept within, the nature, or the common pool. Paradoxes are likely if this expectation does not transpire.	Forestalled/released Recovered Avoided Offset Transferred
Paragain goes to the wider economy for equity and productivity	Emphasis on intersectoral allocation and distribution of wastes and wastages. This fraction passes to 'the wider economy or returns to the 'state'. It might include neighbouring farmers in the same scale/level or upstream or downstream users if going up a scale or level but these neighbourly connections are relatively weak and/or delayed.	Forestalled/released Recovered Avoided Offset Transferred
All four – the parasystem	The balance or emphasis of who gets the gain between the four destinations is of importance. The term 'the parasystem' covers all four destinations.	All five means

environmental terms, water might be saved for storage within a wetland which creates direct environmental benefits.

In addition, a sequence of 'clients' might be seen as another type of destination yet here are subsumed into each of destination. Here 'client' means the final consumer of the goods or services produced in each of the four destinations.

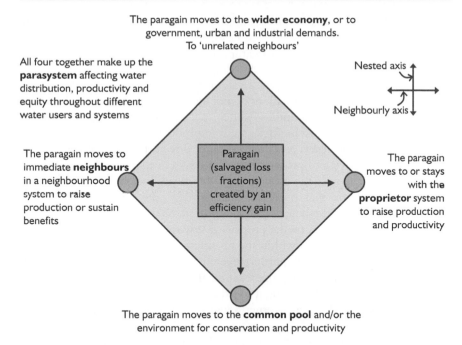

The paragain moves to the **wider economy**, or to government, urban and industrial demands. To 'unrelated neighbours'

All four together make up the **parasystem** affecting water distribution, productivity and equity throughout different water users and systems

Nested axis

Neighbourly axis

The paragain moves to immediate **neighbours** in a neighbourhood system to raise production or sustain benefits

Paragain (salvaged loss fractions) created by an efficiency gain

The paragain moves to or stays with the **proprietor** system to raise production and productivity

The paragain moves to the **common pool** and/or the environment for conservation and productivity

Figure 23 The quadrangle of destinations of resource paragains

For example, a proprietor irrigation system has farmers as 'clients' and those involved in the food chain, including agro-industry and consumers of food.[5] Thus technical water efficiencies achieved by farmers can then be passed, so the theory goes, down the chain to purchasers and consumers.

In addition, the quadrangle is 'fractal'; in other words it exists at different spatial scales. For example, the proprietor system is defined by its boundaries and homogeneity. Therefore a proprietor system covers systems from units such as households and irrigation systems up to cities or river basins. Thus if an efficiency drive is being pursued by a single farmer then her farm is the proprietor system. If a country's whole irrigation sector is the subject of an efficiency policy, then this sector is the proprietor system. These small or large systems compete and cooperate with other neighbouring systems at the same level on the opposite vertex of Figure 23. Further discussion of this scalar dimension is found below.

Furthermore, the four gain destinations arise because the potential salvaged loss exists in the shadow or risk of an increase in scarcity (but not in the middle of a drought). In the shadow of drought, 'pre' a drought, or 'post' a drought, resource users and policy-makers are aware of the need to make savings – and hence there are likely to be salvageable losses. However, within the middle of a drought, the supply of water is so limited that withdrawal and consumption are curtailed. In this situation, losses diminish and therefore cannot be readily salvaged and steered towards the four destinations. Put another way, Montana goes to court against Wyoming's irrigators because of the latter's departure from practice not because there is a drought (Box 1).

It is worth remembering Figure 23 is not about where wastes and wastages go which is why the four destinations do not include drains, landfill and rubbish tips. Reiterating, the paracommons is interested in who gets salvaged resources from systems deemed to be inefficient and unproductive.

Proprietor and single system productivity. Under this purpose, the main focus is the process efficiency and productivity of the system that already is using the resource, in other words the proprietor. To boost productivity and production via efficiency means that more of the 'tare' (in the denominator of the efficiency) 'arrives' in the beneficial good, process or service (the numerator of the ratio) being processed by the proprietor system. Under this focus, savings can be purposively retained within the proprietor system or this retention can work 'implicitly' because of appropriative circumstances at work (see Section 4.5.4). Furthermore, preserving the resource in a storage body (aquifer or dam) belonging to the proprietor, using it to expand areal production or add another crop to the seasonal calender, are other ways that efficiency gains can be captured by the proprietor

As said above, the proprietor system usually includes its 'nested components' or clients – for example, both the irrigation system and its farmers (or the environment and its living ecology, or city and its inhabitants). If this is held to be true, then efficiency savings made by farmers would accrue to the farmer group rather than being passed to bolster environmental flows or to a city's water supply. But this need not be always the case: a closer examination of a farm sitting in an irrigation system tells us that, under some circumstances, it is a fractal of an irrigation system sitting in a river basin. In other words, the farms supplied by a canal are smaller versions of irrigation systems supplied by a river. A fractal view of natural resource systems is then reflected in a fractal view of the quadrangle of Figure 22. To repeat, the proprietor system could be either a farmer who retains the efficiency gain (at the smaller scale) or the whole irrigation system (at the larger scale). At each 'scale' a new quadrangle could be drawn up, resetting the identity of the proprietor, neighbour, common pool and wider encasing 'economy'.

Neighbour distribution, equity and productivity. The aim is to distribute the freed-up tare to users or systems that immediately neighbour the proprietor system. The facet of close neighbourliness forged by efficiency is so important that it is covered in detail in Sections 4.6.1 to 4.6.3 ahead. Briefly, there are two types of connection: parallel neighbours and parasymbiotic neighbours. The former is where an increased efficiency of a proprietor benefits a neighbour where both jointly, in parallel, share a water supply, and the latter is where losses from a proprietor system are immediately picked up by a downstream or later sequenced user if they share a water supply 'in series'. The key definition of a 'neighbour' is that water flows from the proprietor to the neighbour with minimal delays and transaction losses.

Common pool or environmental conservation and productivity. The intention here is to keep the saved resource in the environment or the common pool (CP), achieved by different means but principally by salvaging losses and passing this as an intersectoral allocation (see Molle and Berkoff, 2009, for a critique of this logic). Boosting environmental productivity is predicated upon keeping more of the resource within nature. For example, retaining ecological flows in rivers helps to sustain and secure freshwater biodiversity. For example, if achieved by forestalling losses, this 'tare' would be maintained in the common pool. 'Avoided losses' via capping withdrawal is another way of retaining resources in the common pool. (However, if a downstream freshwater ecology immediately recovers upstream losses, this sets up a much more direct neighbourly connection and is an example of the previous category.)

'Wider economy' distribution, equity and productivity. The aim is to distribute the freed-up gains to users and uses to the wider economy (WE). While this conceivably could happen by all five means, this would ordinarily happen by reducing or avoiding losses and then via a formal route of intersectoral allocation passing this to meet industrial or urban demand. The strength of the connection between proprietor and the CP/WE is reduced by timing delays and transaction losses. The paracommons also lumps the wider economy with the common pool recognising that water might have to flow back to the aquifer or river in order to subsequently be abstracted for another use.

Parasystem distribution, equity and productivity. Here the option is to distribute efficiency gains to an array of destinations – hence the significance given to the 'parasystem' or a system of systems. The parasystem implies that paragains can be shared between the proprietor, the proprietor's neighbours, the common pool, or economy. The parasystem has two axes: neighbourly and nested. The 'x' or neighbourly axis runs between the proprietor and its immediate neighbour(s). The vertical 'y' or nested axis runs between the common pool and wider economy. The 'y' axis implies that savings made by the proprietor could 'flow' (albeit less immediately and less directly) towards a variety of users and uses within the encasing river basin.

Elsewhere in this book, carbon is discussed – with the argument that carbon moves from a carbon dioxide common pool to different ecological services producing carbon dioxide wastage and carbon wastes along the way. In this context, when carbon dioxide is not in short supply, the destinations of 'savings' are more complicated. Picked up in Sections 4.7.3 and 5.3.2, there are two answers, briefly explained here. The first depends on the ecological services sought (for example, long-term sequestered carbon or improvements to flows of catchment streams) which in turn distribute other services (e.g. photosynthesised carbon) towards other resource users in different ways. A second set of 'distribution' outcomes arise from offsetting and transferring carbon services production to other locations and REDD+ type projects.

4.5.4 Appropriative environment : parties/destinations capturing paragains

The context of resource efficiency not only influences interests in efficiency and salvaged resources, it also structures the movement of salvaged resources to one or more destinations. This can be cast as 'appropriative' because paragains tend to be grabbed or held by a favoured party or destination, especially the proprietor. Appropriation can happen either because one water user has a strong or clear capability to use the paragain or because other sectors – by comparison – do not pull the resource gain sufficiently in their direction in turn influenced by mediating factors. The idea of efficiency-predicated appropriation continues Sneddon's view (2007) on resource accumulation via the complexity and materiality of flows and dynamics of nature. Some mediating factors are discussed amongst some of the 20 REC factors, but for ease of explanation, they are collated in Table 15.

Table 15 begins with a number of broad dimensions that set out the context of gain destinations. A greater scarcity of a given resource will drive up interest for appropriating losses by all four destinations. Economic and financial factors also mediate an interest in losses – e.g. a profit motive to make more of a good from a given input means the proprietor will retain and use losses. An ecological rationale to reduce the depletion of the common pool resources will strengthen environmental claims for salvaged losses. Other appropriative mechanisms at work include:

- Property rights. Within irrigation, a water right or licence will usually accommodate or 'correct for' losses during usage. This means the losses belong within the water right which in turn belongs to the proprietor particularly advantaging first in time and first in line rights.
- Parity of property rights. An integrated approach to water management would offer all sectors property rights rather than just those with off-stream abstraction usage. For example, the Tanzanian Water Policy and Laws affords water rights to those users abstracting from a source rather than in-stream users advantaging abstractors.
- Professional norms in irrigation mean that losses are part of the procedure for designing irrigation demands (covered in Section 6.1).
- Infrastructure-sizing. As a consequence of design procedures, irrigation structures are larger than if a smaller efficiency coefficient had been used – this in turn allows irrigators to physically access more water.
- Infrastructure for flow measurement. The measurement of water to different farmers potentially drives down excess use by allowing the metered use of priced water. However, flow measurement for the vast majority of irrigation systems in the world is missing or error-prone which undermines water charging and the regulation of demand (Small and Carruthers, 1991).
- Geography, location and sequencing. The physical placement of water users in relation to others can also structure the distribution of losses: top-enders tend to be more advantaged in accessing water.

Table 15 Appropriative factors directing salvaged losses to destinations

Factor/Property	Driving up (or down) interest on wastes and wastages	Likely paragain destination (Section 4.5.3)
Resource scarcity	Increasing scarcity drives up focus on economising or recovering and reusing waste and wastage.	Increased interest by all four destinations.
Economic growth	Ensuring that resources move to the economically most efficient sector.	To the wider economy (WE).
Financial profit	A profit or commercial motive drives productivity and the improvement of efficiency and recovery of waste/wastage.	Emphasis on proprietor system production and process efficiency.
Green growth priorities	A balance between economic growth and environmental protection.	Proprietor system and conservation of common pool.
Environmental protection	Concerns over environmental harm: environmental flows.	Conservation of common pool (CP).
Property rights	Current regulation mediates who receives the salvaged paragain.	Most likely the proprietor system (e.g. Montana vs. Wyoming).
Regulation and professionalisation of the sector	A greater level of focus on performance originates with professional training and tighter regulation.	Emphasis on proprietor system productivity, production and process efficiency.
Technical and infrastructural advantages and scope	The geography and physical aspects of withdrawal influence the locations of paragains consequently affecting who gets them.	Towards the proprietor system.
Technological capability/ ingenuity	Within proprietor system, a technological change might allow previously unrecovered wastage to be captured.	Towards innovating sector or system. Bearing of costs might also advantage proprietor system.
Pace of change	Slow changes might not be readily discernible.	Relatively small efficiency improvements over time probably favour the proprietor.
Equity and justice	Increasing interests in the distributions of resources between parties.	Emphasis on parasystem distribution and allocation.
Energy-water nexus	Efficiency drives might create inefficiencies in other resources.	Probably paragains to the proprietor, but externalities to the CP and WE.

- Scope for expansion. Losses rendered into salvaged paragains for the proprietor require the potential to expand into new lands by increasing irrigation command area. Although the patterns of irrigated land compared to irrigable land cannot be generally characterised, many farms and irrigation systems can expand into rainfed land.
- Costs. Costs of efficiency improvements usually fall to farmers and system managers with the likelihood that they have a reasonable de facto claim on the gains achieved.
- Justice and parity of cross-sectoral scarcity. During periods of scarcity (temporary or longer-term) as all systems begin to consume less water in response to a reduced supply, the manner in which all fractions of water withdrawal are reduced pro-rata across all users is likely to be unstructured and poorly monitored. This means that sectors with prior claims and larger claims are likely to be favoured.
- Lack of incentives to give up 'losses': for example, top-enders in irrigation systems in Tanzania find it easy to hold onto water depths of about 20–25 cm in their paddy fields (tailenders pond water at about 5 cm) because of the lack of field canals and drains.[6]

Another part of the appropriative environment is that a quanta of a paragain may be extremely difficult to track and send across a landscape to a given destination at a given time. The complexity of water flows through the landscape provides high levels of confusing background noise and a propensity for both appropriation and transaction losses to take place. For example, a saving of, say, 20 litres per second cannot easily be tracked in a catchment pervaded by shifting and overlapping groundwater, rainfall, soil moisture and small streams. In addition, this quanta of water has to be located, relocated and translocated across boundaries such as other abstraction points, farms, administrative boundaries, geologies and so on. The total effect of these factors is the unsurprising conclusion – but paradoxical if our expectation is otherwise – that proprietor systems tend to capture the material benefits of efficiency gains.

4.5.5 Dynamic and recursive context

Part of the complexity of resource efficiency can be ascribed to highly interconnected systems sitting within an unpredictable dynamic and non-static context driven by exogenous factors such as climate and weather (Leach *et al.*, 2010). With reference to directing paragains to one or more of the four destinations, a dynamic and recursive context is most likely to advantage the proprietor system. This is for the reason identified in the previous section: that a high level of background noise will probably make it difficult to extract a paragain from the proprietor system and steer it to other parties and destinations.

Also within agro-ecological systems and external to them, feedback loops create a complex co-emergence of both context driving new efficiency practices and efficiency practices driving context, generating new levels of demand and distributions of resources. Dynamic recursivity captures the effects of changes within one part of the wider system cascading through to other parts in unpredictable ways – and does so that leads to multiple feedback cycles. Hoff (2009), for example, makes the case for distant, global and complex 'teleconnections' between environmental change, water withdrawals and climate/weather cycles. Thus, recursivity is behind the Jevons Paradox in terms of how energy efficiency of the *actual* work or energy gained from a work or energy *potential* translates into 'efficiency-induced consumption of outputs' (Polimeni *et al.*, 2008) by creating more or cheaper work and energy. This phenomenon differs between resources; energy efficiency drives context in different ways from that of irrigation or carbon because the latter two have many material pathways that conversions can follow, whereas energy conversions produce heat, light, electricity, noise and vibration.

4.5.6 The resource efficiency nexus

Resources such as land, water, energy and others, such as nitrogen and carbon, are linked through ecological and agricultural production. Increasing interest in these linkages is expressed via work on the nexus between water, land and energy (SEI, 2011), however, little of this (e.g. Rothausen and Conway, 2011) has examined the role of physical efficiency – though Wagner and Wellmer (2009) address economic efficiency via resource linkages. Changing the efficiency of one resource can impact on other linked resources, adding further complexity (Figure 24). Intensive or extensive use of one resource may have additive or specific intensity effects on another. For example, extensive low yielding rainfed agriculture may be a suitable strategy to save the withdrawal of scarce surface waters for intensive agriculture. An 'opposite' example highlights the possible flaws in attempting to introduce micro-irrigation into rural food production, neglecting the necessity for a reliable and cheap energy supply.

Thus saving water in agriculture by switching from gravity/canal irrigation to pressurised/drip irrigation becomes problematic because water may not have been 'lost' in the previous system, because the extra carbon generated in energy consumption for pumped and filtered water is an unwelcome addition to global carbon, and the extra effort to maintain complex drip systems may worry farmers. In other words, the water-based premise for raising irrigation efficiency exports the higher carbon footprint to the greater or global population. In another example Law and Harmon (2011, p.74) link irrigation and forestry when suggesting that additional forest watering might boost carbon sequestration – yet clearly with consequences for the likely low productivity of that water in competition with other calls on it elsewhere in the river basin.

The role of the efficiency nexus on paragain appropriation cannot be readily divined, and space limits a larger exploration of this topic. However, given that the paracommons and parasystems should be primarily seen from the perspective

Attempts to reduce resource use are interlinked with other resource efficiencies leading to uncertain outcomes for all linked resources (showing water, land and soil, energy and CO_2. Pie-chart segments are indicative).

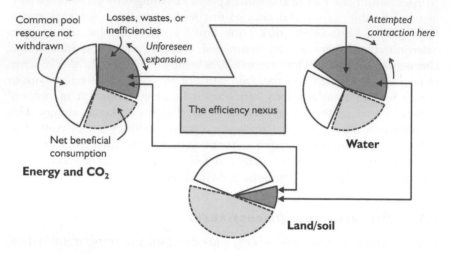

Figure 24 The efficiency nexus (energy, water, land)

of how the proprietor behaves, then one notes that the raising of efficiency of irrigation by offering energy for a pressurised irrigation system plays into the hands of the proprietor. The corollary is that the costs of lower efficiencies in other sectors (e.g. subsidised electricity) are borne by the environment and wider economy.

4.6 Tertiary: scales, nestedness, time/timing and neighbourliness

4.6.1 Scale: spatial, place, boundaries

Perhaps the most important source of resource efficiency complexity lies with the conceptual and practical difficulties that can be sourced to the accommodation of scale, place, time and boundaries. This 'problem' might be best characterised as a clash between our (human) tendency to compartmentalise complex nested systems into zones, units or time windows with defined limits and boundaries and a resource's tendency to flow across and therefore to ignore such artificialities. While this subsection addresses mainly spatial issues and the next discusses mainly temporal issues, space and time are linked by the time it takes for water to travel across catchments and irrigation systems. Below, nine issues are identified.

First, in terms of horizontal scale, knowing inputs, outputs and ratios over a given area (hectares) and time period (days) allows for the step of defining total production and total consumption by controlling for boundaries such as time and area. For example, 18,000 kg of rice is annually produced from a three hectare farm producing 3000 kg/ha with two seasons per year. Furthermore, knowing area, plus the depth equivalent of application (e.g. an irrigation dosage of 950 millimetres over 120 days) allows an efficiency ratio to be converted into a quantifiable volume in cubic metres. Defining system limits allows for portability of results if an efficiency figure is given alongside its areal provenance (e.g. per farm or per total irrigation command area). This element of time period is also connected to ecosystem services where some services are measured in terms of years rather than days or months (for example, long-term versus turnover carbon sequestration). However, unlike a season of crop production the time period for ecosystem services may not be as defined.

Second, vertical scale plays a role in water via elevation and relief, creating (with gravity) a sequence of water abstraction, usage, waste and recovery. This plays out at the micro-scale where a farmer might trap upstream water from her neighbour's drainage or at the basin scale where a wetland depends on drainage flows from an upstream irrigation system.

Third, geographic scale plays an important role in terms of the nesting or nestedness of systems. A farmer's field nested within a tertiary unit nested in an irrigation system in turn nested in an encasing river catchment creates conceptual, accounting and terminological difficulties because local recoverable losses from the field can be returned to the hydrological basin. Therefore, knowing the total amount of water consumed (rather than withdrawn) allows for the impact of an irrigation system on the catchment's water balance to be calculated. Figure 25 shows in simple terms how a field sits within a wider basin. Thus water seeping below a root zone is a loss from the field's point of view, but is not a loss for the basin. This conceptual model allows the difference to be 'accounted' for by the difference between classical irrigation efficiency (CIE) and effective irrigation efficiency. CIE computes that efficiency is the ratio between the beneficial consumption to total withdrawal, while EIE is the ratio of beneficial consumption to total consumption. Thus CIE and EIE are measures of efficiency at the farm and basin levels respectively.

Similarly, energy savings at the household (unit) level may in turn lead to rebound for the broader nesting economy (Sorrell, 2009). In this manner, nestedness and scarcity are associated; the greater the scarcity at the basin scale, the more important it is to account for local losses by referring to what is recovered at the basin scale. Furthermore, defining the scale levels requires care in order to formulate appropriate accounting systems: irrigation sits encased within river basins or aquifers; houses sit within housing communities in a broader economy, and forests are within a global atmosphere.

Fourth, the phenomenon of neighbourliness, important in the paracommons, commences from a spatial awareness that sees one user abutting against another, a connection that strengthens when scarcity increases. In other words, zooming

Withdrawals [W]. Irrigation water withdrawn from basin added to field W = BC + NBC + RF + NRF + ΔS

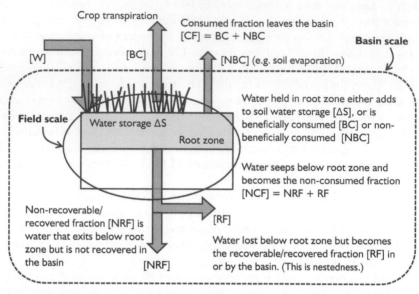

Figure 25 Block model of irrigation; field and basin scales

out sees neighbours coming into the proprietor's frame. This is the reason that in Figure 23, two axes are given: nestedness and neighbourliness. In the former, the proprietor sits within an encasing system, in the latter the proprietor sits alongside a neighbour. To exemplify, Figures 26 and 27 provide a plan and side view of two units side by side. This neighbourly scale can be contrasted with the block or monolithic model of the irrigation system given in Figure 25 where no neighbours are recognised. This distinction between block and bifurcating models of irrigation systems changes the significance of the lost-locally-but recovered-to-the-basin fraction in Figure 25. The block model, the local loss is recovered by the basin. In the bifurcating model this loss is observed by the neighbour as being a genuine loss because he awaits his supply of irrigation that is first being used by his neighbour. In other words, local losses matter to a farmer because those losses matter to his neighbours. This is explained in depth later on.

This means that efficiency becomes interrelational or connective. Thus interrelational efficiency occurs when: one neighbour recovers the losses of another neighbour; the loss fraction in one user subtracts this equivalent amount from a neighbour; or the 'material gain' can be redistributed from one proprietor to its immediate neighbours in the same scalar level or to an encasing (higher scale) environment. The important point is that not only are the users and systems nested, so is neighbourliness. In other words neighbourliness can be direct and immediate within the same scale (one farmer next to another sharing the same canal) but also distant and less immediate (one farmer and a distant

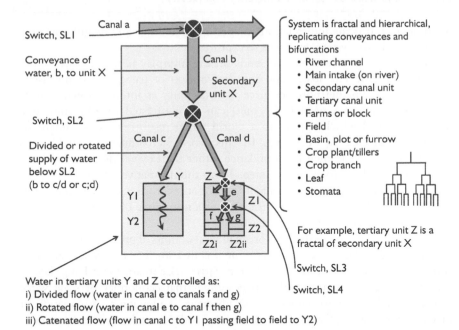

Switch, SL1 ——— Canal a

Conveyance of
water, b, to unit X ———

Canal b

Secondary
unit X

Switch, SL2 ———

Divided or rotated
supply of water
below SL2
(b to c/d or c;d)

Canal c   Canal d

Y   Z

Y1   Z1

Y2   Z2

Z2i   Z2ii

System is fractal and hierarchical,
replicating conveyances and
bifurcations
  • River channel
  • Main intake (on river)
  • Secondary canal unit
  • Tertiary canal unit
  • Farms or block
  • Field
  • Basin, plot or furrow
  • Crop plant/tillers
  • Crop branch
  • Leaf
  • Stomata

For example, tertiary unit Z is a
fractal of secondary unit X

Switch, SL3

Switch, SL4

Water in tertiary units Y and Z controlled as:
i) Divided flow (water in canal e to canals f and g)
ii) Rotated flow (water in canal e to canal f then g)
iii) Catenated flow (flow in canal c to Y1 passing field to field to Y2)

Figure 26 Bifurcating model of irrigation – plan view

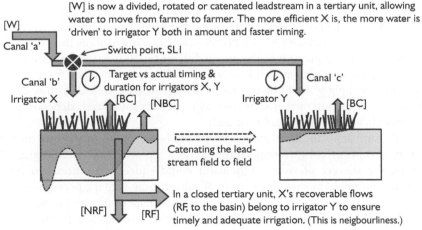

[W] is now a divided, rotated or catenated leadstream in a tertiary unit, allowing
water to move from farmer to farmer. The more efficient X is, the more water is
'driven' to irrigator Y both in amount and faster timing.

[W]
Canal 'a'   Switch point, SL1

Canal 'b'   Target vs actual timing &   Canal 'c'
                       duration for irrigators X, Y
Irrigator X   Irrigator Y
            [BC]   [NBC]   [BC]

                    Catenating the lead-
                    stream field to field

                    In a closed tertiary unit, X's recoverable flows
[NRF]   [RF]   (RF, to the basin) belong to irrigator Y to ensure
                    timely and adequate irrigation. (This is neigbourliness.)

Division: Irrigators X and Y share main supply [W] by dividing canal 'a' water at SL1 to canals 'b'
    and 'c' simultaneously.
Rotation: Irrigator X irrigates with whole of [W] water and takes time doing so. Then X closes
    canal 'b' at SL1 and passes whole leadstream to canal 'c' for irrigator Y.
Catenation: If SL1 and canal 'c' are not present, irrigator X uses whole of [W] water and after
    completion passes this flow to irrigator X as surface flow in field to field irrigation.
    Water continuously flows to canal 'b' to supply both irrigators.

Figure 27 Bifurcating model of irrigation – side view

town or city). It is the interrelations shaped by nestedness and neighbourliness that the term 'parasystem' attempts to capture.

Efficiency as interrelational depends on the interplay between scarcity, the materiality of the tare losses and the distribution of paragains, as discussed in Section 4.5. Furthermore resource use efficiency as interrelational (where neighbours' concerns need to be met) can be contrasted to efficiency as a 'normative' goal which applies to a world when resources are abundant or not interconnected by recycling. In this world, a competing neighbour's resource needs are framed by abundance rather than by seeking to recycle or reduce her neighbour's waste/wastage. Thus, the normative goal (the 'ought') to improve efficiency in an abundant world comes from a sector-specific and professionalised discourse about efficiency as a normative object to address via benchmarks and industry-set standards.

Fifth, is the question of the size (or smallness) of units of production contained within a system. Thus a common argument is that small farms are less efficient and productive than large farms or factory farms (Bachman, 1952; Lang, 2003). Although evidence is mixed, the suggestion is that economies of scale work to the larger farmer's (or larger field's) advantage – allowing a combination of mechanisation, seeds, fertilisers, water control to be brought to bear upon crop production. This topic generates heated discussion because of its view, so the argument goes, that the interstices of nature and human life or of niche methods, products and marketing (an example being organic production and produce) have no place in 'efficient' factory farming. In the paracommons view of efficiency, I am able to partly parenthesise this thorny cultural question because I am interested in the technical efficiencies of natural resources rather than the styles and qualities of production systems. Nevertheless, one might argue that size, scale and neighbourliness play out differently in different circumstances and cultures – that large fields and farms as well as very small fields and farms allow for different ways of controlling water to be employed (long furrows and small drip systems, respectively) which in turn give rise to varying performance levels depending on how well they are designed and managed.

Sixth, similar to the fifth point and arising out of it, are questions on the density of units offering an opportunity to set a cap on the 'per unit demand'. Raising the density of units of demand could squeeze out non-beneficial demand, and might be seen as a version of substitution. Like the previous point, generalisations are not easy to develop given that smaller units within the same area could play out either way – leading to less or more efficiency. Nonetheless, the debate in North America on urban smart growth is predicated on higher density housing that generate both lower net demand per unit as well as more compact services delivery and higher efficiencies, though with a trade-off of fewer 'green spaces' (Danielsen et al., 1999; Greca et al., 2011).

Seventh, understanding scales, nestedness and time allows us to define the boundaries, and therefore the identities, of complex systems still producing recycling wastes and pollutants. This is central to determining how well we manage social-ecological systems, simultaneously knowing how transparently we control externalities such as pollution that harm natural systems and productive ability. Boundary control is significant when humans seek to change or obfuscate boundaries (for example, by extending time, neglecting sinks or by borrowing) for the purposes of demonstrating higher performance than otherwise possible without better coordination of inputs. Boundary determination is also important for adjudicating decisions on water rights (Skaggs *et al.*, 2011) and to derive a more complete picture of resource accounting in order to judge performance more holistically – given as society generally attempts to correct free-riders seeking to more cheaply export pollution to nature or marginalised members of society. These two points further appositely inform the performance debate when in a globalising and more inclusive view of an interconnected planetary commons, obfuscating boundaries is increasingly less tenable or acceptable. Furthermore, the point I follow up on later is that it takes resource recycling at nested scales for society to understand that once previously externalised pollutants and wastes are now internalised within a larger system of recycling.

Eighth, is 'scalar reach', the extent to which one proprietor system can affect or reach out to other systems and the wider commons depending on how an 'inefficient' and large the proprietor irrigation system is. This reach is proportionally related to the inefficiency intensity: the more water 'wasted' within a system the greater this shapes and skews current distributions and the greater the 'tare' or extrinsic fraction for future distribution. The changes in efficiency that derive the 'delta tare' or paragain are also proportional to the starting position of inefficiency. Thus reach also comes from the hoped-for distribution of these gains amongst the various sub-systems. As explored later, this current and future 'wasted' fraction, real, legally attributed or politically imagined, impinges not only on downstream users but also on upstream users of water. It has an upstream effect because the inefficient fraction could be legally argued to be part of the downstream system's 'need' and thus an upstream irrigator has to 'sit and watch' while a flow passes her intake for a legally ascribed but inefficient downstream system. Salman (2010) covers aspects of upstream reach from legal downstream use. This wider reach, explained by the coined term 'parageoplasia', occurs when the proprietor system drives changes remotely.

4.6.2 Scale: time and timing

As mentioned above, scale and place are connected to time and timing. It does this because many resources cannot arrive instantaneously at different places to meet all users' needs simultaneously. Instead resources flow, taking time. In addition, timing is important if we are interested in crops or living entities with production curves influenced by the timely arrival of a resource input. The degree to which resource users and managers fall to control for time and timing offers another uncertainty element of REC. Time and time boundaries are also critical to accounts and accounting frames. Below, eight dimensions of time are briefly introduced.

The first aspect is where pathways have a flow-timing characteristic, e.g. water flows quickly through a stream but slowly over and through soil. Thus, 'losses' to one part of the system may, if given enough time, be recoverable to the same system or to the next scale up. However, despite this water being recovered by the aquifer and or catchment, it moves at a much slower pace than if that water was retained in the surface water network. Thus, recovery may be beneficial in terms of higher system efficiency, but reduces productivity if the delayed arrival of downstream water harms crop or ecological life (Lankford 2006, 2012).

Related, is a second aspect of time scheduling: that local field losses, recovered to the basin, slow down the rate of irrigation progress leading to delayed scheduling and greater incidence of water stressed crops. This strengthens the physical connections between neighbours and is explained in more detail in Section 4.6.3.

Third it becomes important to define the time boundaries for a system under study in terms of days, months or years. By way of example, water used efficiently and productively within a cropping season of an irrigated crop may sit as open water in a field after the crop is harvested. If the time horizon of study is expanded beyond the harvest point, this standing water is lost by non-beneficial evaporation. The irrigation efficiency drops if post-harvest evaporation losses are taken into account.

Fourth, stemming from the previous two points regarding delayed timing and time windows is an accounting problem: the decision over whether to put these attenuated flows inside the 'season' or time window being analysed and accounted, or whether they fall within the next time window and therefore are effectively depleted from the first time window.

The fifth point about time arises when an economic or livelihood rebound effect takes place because demand increases as a result of 'time saving' for a process or group of people (Hertwich, 2005; Ruzzenenti and Basosi, 2008). A well-known example of this is where domestic household water demand is inversely correlated to the distance and time required to fetch and collect water.

The sixth aspect relates to shifts in efficiency improvement between two times, for example, between the present and the future or between what was designed/built and the present. This observation follows the various ways efficiency can be discussed as 'futures'. This can happen in absolute terms (i.e. 'that irrigation systems should be "xy" per cent efficient'), or comparatively ('policies are required to make irrigation systems more efficient'). These utopian efficiency futures, expressed either without a timeline or by using a project time-line of four to five years, are problematic. Much of the paracommons hints at the multiple outcomes that systems can decompose to; they imply that future alternatives will differ from each other at different scales and will differ paradoxically from an expected future.

Seventh, from the previous point, a resource gain comes from the differences in efficiency between the starting point (time 1) and end point (time 2). But this paragain will be understood differently depending on circumstances and depending on the definition of the starting and end point. Table 16 provides a basic framework for how five starting and three end points might be determined and how calculating different amounts of paragains arises from selecting different windows (and there may be other start and end points). This issue again underscores why the term paragain has been chosen as the term for salvaged resources suggesting a high incidence of subjectivity and assumption.

Table 16 Different start and end points for comparing efficiency over time

Starting point – the 'today'	Explanation
Design and build	Assumed or designed parameters for efficiency and behaviour are taken from the design specifications. If design assumed very low efficiency then expected gains might be higher than otherwise foreseen.
Average climate; past 'x' years	Records from the last five to ten years might comprise either real or assumed parameters. For example, if the previous five years were reasonably 'wet', and the future time window falls in a dry period, then paragains cannot easily be computed without correcting for moisture differences.
Last year's events	More clearly etched in the memories of farmers, engineers and policy-makers, the 'today' starting point significantly draws upon recent events such as a drought.
Now, this year	With sufficient measurement, system managers might track current water use and discern current levels of efficiency and losses.
Assumed now	Probably the most likely scenario is to guess the current status of the system without reference to measurement, design or recent behaviours.

End point – the 'tomorrow'	Explanation
Expected future	Policy intended future. This is the prefigured future, without reference to detailed and accurate measurement.
Actual future – end of this season	With sufficient measurement, tracking of losses and efficiency might be possible contemporaneously.
Actual future – a point 'x' years in the future	Again with measurement, it should be possible to generate a post-project evaluation of impacts of efficiency improvement programmes.
Actual average future; average future 'x' years	Ideally impacts should be assessed on the average of a number of years allowing climate variability and other changes to be corrected for.

Lastly, if the time window is too long then the relative gain may be deemed insufficient. This is akin to saying that society should not be surprised if a 5 per cent gain in salvaged resource derived over five years stays with the proprietor system, a different prospect from a 5 per cent gain being achieved in, say, three months. The longer time horizon allows for obfuscation or 'mission creep', to use a popular term. A shorter more abrupt time horizon could make the gain more noticeable.

4.6.3 Parasystem distribution; neighbourliness revisited

The juxtaposition of the previous sections on context, materiality, destinations and scale/place/time gives rise to a) the issue of the destinations and distributions of waste/wastage fraction to neighbouring parties receiving or controlling those fractions and b) different kinds of interrelations between those parties. Dissecting neighbourliness is critical to understanding how changes to efficiency impact on 'third parties' (Pease, 2012). This section delves into these interrelations in more detail and deals with two parts of the efficiency quadrangle in Figure 23; 'neighbourhood systems' and the whole quadrangle termed the 'parasystem'. This topic arises because of particular scalar and neighbourly dimensions; for this reason the REC framework places this topic in the tertiary group of complexity sources dealing with the accommodation of scale, boundaries and time in Figure 19.

Neighbourhood systems – closely connected neighbours

There are two types of closely connected neighbours – one type connecting losses between neighbours, the other where downstream neighbours receive losses from upstream irrigators. Employing words with the prefix 'para' allows me to classify the former as parallel neighbours and the latter as parasymbiotic neighbours. Table 17 and Figure 28 help explain this approach though in the top row of Table 17, for comparative purposes, I include the example of the proprietor system without neighbours. Here a farm or irrigation system expands its command area by a combination of reducing and recovering losses. Thus the proprietor system has retained the losses for itself. Referring to Figure 28, Farm A has expanded to a new command area that includes the area Aa.[7]

PARALLEL NEIGHBOURS

Parallel neighbours are physically connected by sharing water and 'losses' – and therefore by the benefits of reducing losses (Lankford, 2012b). Parallel neighbours witness their water management being partly a function of how an individual farmer operates her irrigated farm and partly a function of how her neighbours operate their farms. Thus efficiency improvements being made by all or the majority of the combined farms affect the scheduling of water between all farms. In the middle two rows of Table 17 two types of parallel neighbours

Table 17 Types of neighbours closely connected to the proprietor system

System and means	Destination example and effect on parasystem productivity
Proprietor only Forestalled/reused losses Recovered/reused losses	*Proprietor irrigator only.* The original irrigator (A) extends her command area to Aa by forestalling or recovering the extrinsic fraction and converting this to beneficial consumption. Production for proprietor A is raised by increasing area but this impacts on other users if recoverable flows diminish. Production trade-off is mixed.
Parallel neighbours connected by forestalled losses	*Rotational-flow irrigators.* Both proprietor irrigator and neighbouring irrigators benefit from more rapid cycling of water by reducing field level losses. Crop production within the irrigation system is raised by more rapid and frequent irrigation scheduling.
	Divided flow irrigators. Both the proprietor irrigator (A) and the neighbouring irrigator (B) benefit from ensuring a more equal division of water to the command area by correcting any losses within one of the bifurcations. Irrigator (B) raises production by being given more water not lost by (A).
Parasymbiotic subordinate neighbours connected forestalled losses and by recovered losses	*Catenated flow irrigators.* The downslope subordinate neighbour (C) benefits from a more rapid delivery of his neighbouring irrigator's (B) water because irrigator (B) upstream has forestalled and released her 'losses'. Downslope production increases due to earlier timing of water.
	Downslope subordinate irrigators. An irrigator (D) in a low spot or hollow grows a crop using drainage water from irrigators at a higher elevation. A high water table has the same effect. This irrigator benefits from greater 'losses' created upstream.

Notes: The location of irrigators A to D are shown in Figure 28

are explored. These are: rotational-flow irrigators (where irrigators are on a rotational supply and take water in turns) and divided flow irrigators (where irrigators split a flow and each takes a continuous supply). Thus in networks of dividing canals whereby recoverable losses in one branch of a division can not be counted upon to arrive at the other branch of the division point it becomes more relevant to forestall losses in one bifurcation so that a rotating leadstream more quickly passes to the other farmers in the same or other bifurcation.

Of the two types of parallel connection, the rotated flow could be thought of as the most 'interesting' because here a single inefficient farmer ends up harming his own crop production when his water losses slow down progress for all farmers in the rotational sequence and delays the return of the leadstream (main d'eau) back to him. In the divided flow case (and the catenated example below), a single inefficient irrigator simply 'exports' these loss-induced delays to farmers on the other side of the flow division or further down the chain of irrigators.

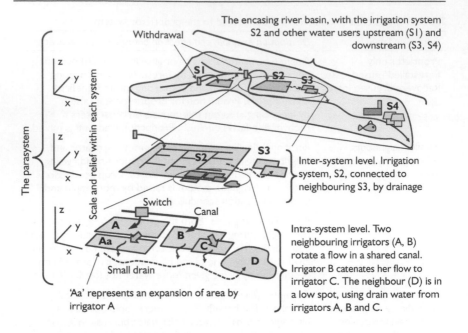

Figure 28 Neighbours, nestedness and the parasystem

PARASYMBIOTIC NEIGHBOURS

There are two types of parasymbiotic neighbours. First, catenated flow irrigators (where irrigators are on a slope in a sequence) take turns one after another, but unlike the rotated flow, water does not cycle back to the beginning. In Figure 28, this is farmer B to farmer C. Second, in the bottom row of Table 17, subordinate irrigators recover the losses from upstream. In Figure 28, a downstream neighbour D recovers the losses expelled by the upstream neighbours A, B and C. The differences between these two examples are instructive; in the previous catenated connection, the upstream farmer has to be less wasteful in order to forestall losses so that Farmer C receives more water more quickly. In the second case, Farmer D depends on Farmers A, B and C being more wasteful (if their waste is in the form of recoverable flows).

FORESTALLING LOSSES BETWEEN PROPRIETORS AND NEIGHBOURS

In the following text, I explain how forestalling and releasing losses can benefit neighbours. First we have to return to the normative idea of irrigation efficiency; that a high standard of water control is required to minimise losses. To explain this concept requires a simple contrast between two types of irrigation scheduling given in Figure 29. Both graphs show time along the x axis and soil moisture deficit along the y axis. The aim of an irrigator is to schedule an irrigation dose

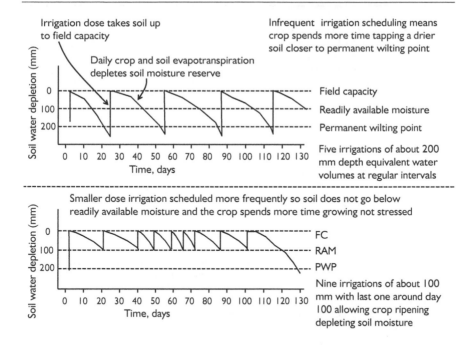

Figure 29 Control of water for irrigation scheduling

'on time' prior to the depletion of soil water moisture. However, soil water is held in the soil matrix at different tensions at different volumes; there is readily available moisture (RAM) between field capacity and about 15 to 25 per cent volume, and then less readily held moisture between RAM and permanent wilting point (PWP). The message here is about timing – if water is delayed, crops grow in drier soil, impairing growth.

So I now connect losses to timing. In Lankford (2012), I described the physical connection between farmers' fields that share a fixed supply of water that flows to a fixed command area. This connection arises because of the continuity equation that relates time to losses. The equation is:

(applied dose of irrigation, mm depth) =
  (flow rate, l/sec) × (time taken to irrigate, hours) ×
  (classical efficiency as '*x*'%) × (0.36)/(completed area irrigated, hectares).

The equation says that of the water destined for an irrigator's field, the more that is lost locally (even if picked up by the river basin), the slower the irrigation scheduling. This is shown pictographically in Figure 30 where losses below the root zone do not contribute to completing the field on time. This makes inefficient systems (in classical terms) poor time keepers, even if the effective irrigation efficiency of the system is high. On this basis, one can now argue that collectives of farmers irrigating from the same supply system collectively

These water losses [NRF + RF] below the root zone are now not available to this particular field to achieve the correct advance rate in order to irrigate on time. Recoverable losses become real losses in the socialised localised efficiency model

The field's 'design' – its soil type, furrow shape, length, gradient, roughness, dryness, density of weeds etc., affect friction losses and influence for a given inflow the infiltration rate, seepage rate and rate of completion

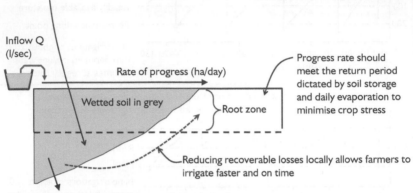

Inflow Q (l/sec)

Rate of progress (ha/day)

Progress rate should meet the return period dictated by soil storage and daily evaporation to minimise crop stress

Wetted soil in grey

Root zone

Reducing recoverable losses locally allows farmers to irrigate faster and on time

The recovered fraction [RF] goes to the basin further downstream, so not seen as a loss in the basin-allocation model of effective efficiency

Figure 30 Efficiency, losses and timing – single field irrigation progress

gain when all of them reduce local losses. To repeat, this is because more water remains in the soil/root zone, allowing for an earlier completion of a field in a schedule of fields. Improved time keeping for the arrival and completion of irrigation ensures that crops grow more productively in soils kept moist and with less water stress (the lower part of Figure 29).

The continuity equation in the previous paragraph and Figures 27 and 30 are employed to create a worked example of how local losses connect neighbouring farms on a canal system. Figure 31 shows a collection of four farms making up a tertiary block of 19.5 hectares sharing a single canal. On the left-hand side, with a classical irrigation efficiency of 55 per cent, it takes 10.94 days to complete the cycle of irrigation with a leadstream flow of 30 litres per second irrigating 24 hours per day. By raising the efficiency to 80 per cent, keeping more of the previously 'lost' water within the root zone, the cycle is reduced to 7.52 days with crops being watered more regularly and prior to the depletion of readily available moisture (as seen in the bottom part of Figure 29). Added to Figure 31 are the sums of total cubic metres entering and exiting the system, the latter by transpiration and drainage (assuming this easy division for the purposes of the illustration). The example shows that a higher classical efficiency of 80 per cent means the whole tertiary block uses less time to complete and therefore less water is withdrawn, and more of the water that is used stays within the soil reserve which means the individual farmers complete their irrigation more rapidly.

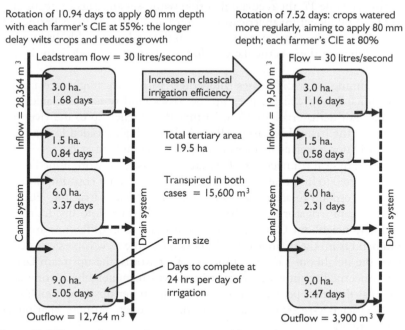

Figure 31 Efficiency, losses and timing – grouped farmers' irrigation progress

Means of salvaging losses and effects on neighbours

Table 17 can also be viewed in terms of how losses may be salvaged by being forestalled, recovered and avoided. For the sake of clarity, I have omitted the other two of the five means (offset and transferred) because they exist as ways of adjusting consumption and losses between users by using one or more of the main three ways already mentioned. Of the three options, I shall first deal with forestalled losses followed by recovered and avoided losses.

With respect to 'forestalled and released losses', a proprietor irrigator makes savings and uses them to expand his or her irrigation area. This is the top row of Table 17. An example is where land-levelling takes place giving a much more even depth of water applied over the field, allowing in turn a lower dosage (without land-levelling a higher dosage may be required to deal with dry/high spots). In Figure 28, the outcome is that Farmer A expands to area Aa. This would be to detriment of Farmer D sitting in the low spot. In other examples, Farmer A forestalls/reduces his losses, and releases them to Farmer B who now receives his irrigation sooner than expected. Farmer B can also pass on unused forestalled losses to Farmer C.

In Table 17, two instances of recovery are described. Losses may be recovered to the irrigator making the losses – in other words the proprietor irrigator – for example, by pumping water from a drain. Second, in the very bottom row of Table 17, the recovered option witnesses an irrigator occupying a low spot or hollow within an irrigation system. In Figure 28 this is Farmer D. Significantly, he or she gains more water when the upslope neighbours are less efficient – although there is the danger that this farmer is flooded out by too much water.

Finally, aiming to avoid losses altogether, Farmer A might reduce his or her command area which passes up the losses (and water associated with retrenchment or dematerialisation) to Farmer B or higher up to the river and other users in the basin. Like 'forestalling and releasing', avoided losses mean that Farmer D now receives smaller drainage flows from Farmer A.

As well as physical connections, there are social issues at work. One farmer, socially connected to his or her neighbour in a queue of water consumption and timings, may respond to accusations of being an inefficient irrigator. The fact that the upstream irrigator B puts basin-recoverable water into his drain may not at all help neighbour C if the ground and water level difference are such that command from the drain water is impossible or the flow routes too slowly to irrigate the neighbour's crop in time. In other words, the upstream farmer becomes concerned about losses because his downstream neighbour complains of delays that the latter perceives as originating from her upstream neighbour. This is a efficiency-related social connection between farmers sharing an irrigation network.

Summary: the parasystem and intersectoral allocation

A 'parasystem' arises from the bringing together of users connected by efficiency and recycling. The two scales of a parasystem are neighbourliness (within the same level) and nestedness (across levels where distance, other flows and transactions losses make it difficult to discern the directness of the connections between parties). Figure 32 depicts a parasystem expressed at these two types of scales and reveals the four destinations for efficiency paragains. A change to the proprietor's efficiency cascades material changes to: 1) the proprietor system; 2) parallel neighbours within the system (e.g. smallholders) or subordinate neighbours next to the proprietor; 3) the common pool, in this case the river, and 4) other subsystems located both upstream and downstream to the proprietor system that make up the wider encasing system.

Summarising this section, brief conjectures about the complexity of efficiency-premised intersectoral allocation may now be made:

- Efficiency-premised efforts to offset efficiency and transfer production to systems elsewhere in the river basin result in redistribution and establish new uncertainties regarding when and where resources, wastes and wastages are subsequently and ultimately consumed and produced.
- Inter-sectoral allocation is unlikely to arise distinctively from different ways of 'salvaging' water for donation to another sector. Rather efficiency-

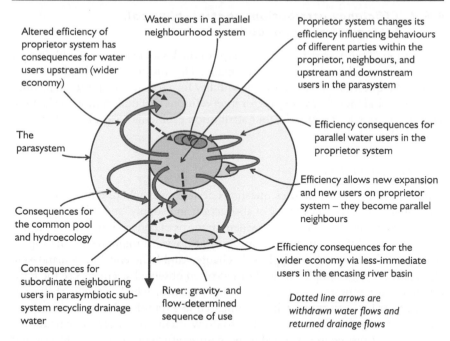

Altered efficiency of proprietor system has consequences for water users upstream (wider economy)

Water users in a parallel neighbourhood system

Proprietor system changes its efficiency influencing behaviours of different parties within the proprietor, neighbours, and upstream and downstream users in the parasystem

The parasystem

Efficiency consequences for parallel water users in the proprietor system

Efficiency allows new expansion and new users on proprietor system – they become parallel neighbours

Consequences for the common pool and hydroecology

Consequences for subordinate neighbouring users in parasymbiotic sub-system recycling drainage water

River: gravity- and flow-determined sequence of use

Efficiency consequences for the wider economy via less-immediate users in the encasing river basin

Dotted line arrows are withdrawn water flows and returned drainage flows

Figure 32   The parasystem of destinations and parties connected by efficiency

premised allocation is more likely to occur via a mix of approaches (recovery, forestalling and avoidance) cumulatively adding up to provide a meaningful supply to another sector.

- The interplay of the two axes of neighbourliness and nestedness produce opportunities for efficiency-premised allocation to be thwarted because salvaged water (the paragain) generated in the proprietor system can be used by the latter or its neighbour. Thus this axis, comprising both proprietor and neighbour, competes with the nested axis, comprising the common pool/wider economy, over this freed up resource. Under these circumstances, the allocation of efficiency savings from the proprietor system to the common pool or other sectors could become even more problematic and tenuous than previously considered (e.g. by Berkoff and Molle, 2009, and Perry, 2007).

- Recovery of waste downstream supplies downstream neighbours with water – but not upstream (or not earlier in the sequence). A greater volume of recoverable waste within the donating proprietor system leads to more waste recovered by the downstream sector or user, all else being equal. Because of the continuity of water flow in catchments, these downstream effects can be immediate or distant and less distinct in large river basins.

4.6.4 Efficiency attribution; spatial, physical, biological and economic

Attribution of the precise location and type of the losses in a chain of withdrawal, consumption and production is problematic. The question of attribution brings to fore the question of who has responsibility for reducing aggregate depletion, who gains, and therefore who pays for interventions that underpin such changes. Using irrigation as an example, four attribution problems arise: physical, spatial, biological and economic.

Physical attribution

Physical attribution establishes questions arising over where efficiency sits; in other words, whether 'efficiency of abstraction and supply' and 'efficiency in the system of use and demand' are equivalent in terms of conversion and complexity. In this debate, different nuances of meaning also arise; it might be useful to apply the term 'effectiveness' to supply-side changes and 'efficiency' to demand-side alterations. [8,9] For example, the author has often observed scientists use the term efficiency when describing the extent and ease with which water is abstracted from a river by a concrete intake as compared to a traditional intake constructed from local materials. Princen (2003) also drew attention to normative neutrality in the lack of distinction afforded to harvesting efficiency that saves forests and harvesting efficiency that would clear-fell forests.

In the original Coal Question posed by Jevons in 1865, as I interpret it, he was vexed by the effectiveness of converting coal to power and heat (in other words efficiency on the supply side) rather than efficiency on the demand side in factories and households – though the parallels between supply and demand in the continuum of consumption and production were rapidly made (Polemini *et al.*, 2008). In Figure 37, in the next section, the efficiency or effectiveness ratio of withdrawal from the common pool is shown on the left-hand side of the diagram, though not without other caveats withstanding such as how much of the common pool is potentially available to withdraw. In some respects, while these definitions and computations are different, the material losses and efficiency gains in both the supply and demand chain cumulatively contribute and pool together – making their distinction for the redistributive and appropriative elements of the paracommons more complicated.

Another physical attribution problem is the tracking of the precise conversion pathway of one key input with losses in amongst a 'background noise' of other inputs – most likely varying in place, timing and intensity. For example, water from a varying shallow water table or small springs feeding parts of the irrigation command area adds to the 'noise' of other input flows. In addition, multiple rain gauges are required to isolate what crop growth evaporation is attributable to main canal inflows rather than rainfall. (My interpretation of literature modelling of sequestration within the carbon cycle is that the same attribution problem exists due to the high number of coexisting inflows and outflows.)

Spatial attribution

Irrigation managers struggle to distinguish water losses taking place within the main canal system (particularly if the whole system is owned by farmers) from losses within farmers' fields. The main canal is a single stretch of water yet often deemed to be owned collectively or by the government. Fields, on the other hand, are owned individually, but give rise to collective grouped efficiency at the tertiary or secondary level. Thus with regards to losses from the main canal there is a problem of ascribing collective responsibility of reducing losses to farmers. In addition, farmers may quarrel over how main canal losses compare with the individual yet collective losses from their individual fields. (Note this distinction is made for the purposes of illustrating the complexity of attribution; I would not equate the main canal as the supply-side and farmers' fields as the demand-side. One might argue that all sits on the demand side, being below the abstraction point from the river – a distinction posed by FAO, 2012.)

In addition, one must recall that these losses may be recovered by neighbouring irrigators or other basin water users – thus a savings programme would generate unknown and disproportionate gains and losses for all involved. All of these make a formal irrigation efficiency programme or intervention difficult to design, cost, pay for and maintain. Thus without a sufficient density of measurements we are often unable to know where geographically losses arise.

Biological/agronomic attribution

In the fractions literature (Perry, 2007), withdrawn waters are divided into categories such as beneficial fraction and non-beneficial fraction. While this is useful in some respects for conveying the results of water accounting, in reality, waters cannot be divided so easily. As well as the other attribution problems listed here, waters are divided into a continuum of productivities. Figure 33 reveals a theoretical crop yield response to water applied showing at the left-hand side a very sharp response to small amounts of water and on the right-hand side a declining response to excessive watering. While the curve reveals multiple points on this response curve, five categories are also given. These range from protective irrigation (small amounts of water applied during a drought) to ensure some yield is achieved, to deficit irrigation (acceptably stressing the crop thus reducing the amount of water consumed but ensuring a satisfactory yield) to full irrigation (ensuring maximum returns to land).

Economic attribution

Also obscure is whether investments in efficiency are adequately recompensed through benefits and resources saved. In irrigation the costs and economics of loss attribution might be addressed through on-going maintenance paid for by farmer levies (a notorious weak-point in system management) or by periodic rehabilitation, usually paid for by external donors (see Mateos et al., 2010, for example).

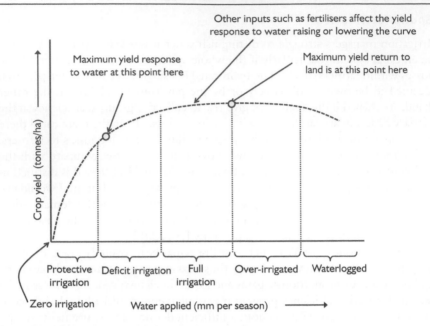

Figure 33 Idealised crop yield response to water

4.6.5 Control, switching and coupling: technologies of efficiency improvement

Sitting between physical resource inflows and outflows to produced goods and wastes/wastages sit technological social-ecological systems commonly nested within other systems. The technological character of these systems – particularly irrigation systems – can often be interpreted as a 'black box' implying lack of detailed knowledge of cause and effect. This is understandable. Although scientists and resource users might physically observe where resources begin (as river water) and 'end up' (e.g. in a drain or as a recovered fraction), the means and ability to control water through systems towards chosen end-points is by no means simple and transparent. The multiple, diffuse and inexact means to control and switch resources generate significant uncertainties.

In this discussion, I employ 'water control' as an apportionative term, rather than political sense of 'taking control of water' e.g. Boelens (2008). Apportionative water control has been explored by the World Bank (Plusquellec, 1994) and FAO (Renault *et al.*, 2007) in their irrigation studies but especially defined by Bolding *et al.* (1995) where the authors defined 'water control' as central to the political economy of water distribution. Nevertheless, questions arise over whether poor apportionative control plays into the hands of those wishing to 'take control' over water (Zwarteveen, 2008, referring to research by Oorthuizen). There are seven dimensions of resource control germane to resource efficiency complexity and the paracommons.

First, is a concept that sees the efficiency ratio as not necessarily only a performance measure or outcome of what has taken place but as a model of how to manage water. In other words, irrigation efficiency tells us that we require a level of control so that losses are minimised. This is the normative and non-prejudicial interpretation of efficiency mentioned previously. To reinforce this idea, one can show how important it is to have field control of water on a sloping field in order to properly water and wet up the soil profile without excess water draining off at the end of the field (Figure 34). While this drainage water might be picked up a couple of kilometres away once 'command' has been re-established, this water is not easily returned to the neighbourhood of farmers sharing a canal because the drain is below the level of the fields (see Figure 14).

Second, there is a large array of measures one can apply to improve water control, raise efficiency and reduce losses. Table 18 has arranged a sample of measures using a framework of three main objectives; i) reduce duration of irrigation need; ii) control of irrigated command area; iii) reduce specific water demand (also defined as the hydromodule). The third objective is examined in Table 18 by classifying actions between the start/end and middle of the cropping season, and by dividing irrigation in the middle part of the season into field level control and canal and gate level control. Each of the actions requires other matters to be considered that are not in Table 18 – for example, the bylaws and sanctions currently in place to strengthen these decisions.

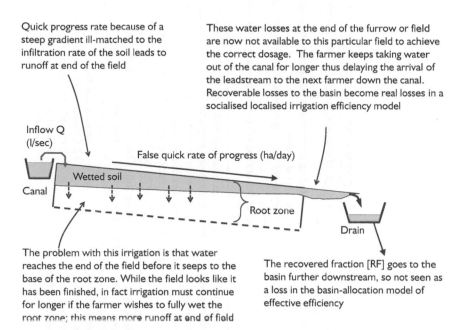

Quick progress rate because of a steep gradient ill-matched to the infiltration rate of the soil leads to runoff at end of the field

These water losses at the end of the furrow or field are now not available to this particular field to achieve the correct dosage. The farmer keeps taking water out of the canal for longer thus delaying the arrival of the leadstream to the next farmer down the canal. Recoverable losses to the basin become real losses in a socialised localised irrigation efficiency model

Inflow Q (l/sec)

False quick rate of progress (ha/day)

Wetted soil

Canal

Root zone

Drain

The problem with this irrigation is that water reaches the end of the field before it seeps to the base of the root zone. While the field looks like it has been finished, in fact irrigation must continue for longer if the farmer wishes to fully wet the root zone; this means more runoff at end of field

The recovered fraction [RF] goes to the basin further downstream, so not seen as a loss in the basin-allocation model of effective efficiency

Figure 34 Efficiency, losses and timing on sloping fields

Table 18 Measures for improving water control and reducing irrigation need

i) Reduce duration of irrigation need	1. Season length of crop variety (e.g. in Tanzania some rice varieties grow for 150 days, some basmati types mature in 120 days). 2. Field wetted up at start of season leading to longer evaporation: in Tanzania some irrigators utilised 30 days, others 7 days, tailenders used 2 days. 3. Field drying-off and irrigation ceases approximately 7 to 30 days before the end of the season depending on crop, soil and climate. (Two other factors extend period of water in the fields; ratoon cropping of rice and use of mixed or unclean rice seed – see also below).
ii) Control of the irrigated command area	4. Tighten up areas watered but not planted e.g. field and rice nursery boundaries, accidental spillage, etc. 5. Reduce accidental irrigation into harvested plots (commonly found in top-end fields). 6. Close down abandoned areas totally or cease irrigating wet areas that receive water from a high water table. 7. Reduce area of late-planted rice cropping likely to produce very low yields due to winter temperatures. 8. Reduce or relocate area of high altitude irrigated crops where agreement is they probably don't need irrigating. 9. Extend duration of rice nurseries so growth takes place here where water control is better and area is smaller.
iii) Reduce specific water demand (also defined as the unit demand, or hydromodule) or improve uniformity of	a) Beginning and end of irrigation season	10. Reduce amount of water in wetting up dry soil (some farmers used two doses of 250–300 mm; others 150 mm). 11. Locate rice nurseries and non-rice crops close to main intake to reduce travel time and possible seepage losses. 12. Channel water to distant fields and nurseries by small earth canals rather than by open field-to-field cascading of water. 13. Band or zone the transplanting of rice so all bunded paddy basins are irrigated/harvested in one area simultaneously. 14. Use clean uniform seed (allows uniform ripening and earlier cessation of irrigation).

hydromodule over irrigation system	b) Mid wet season and throughout the crop season	*Field scale and irrigation scheduling* 15. Control the depth of standing water layer in rice plots (ideally between 5–10 cm while some farmers had more than 20 cm). 16. Change size of bunded plots/basins – smaller when land is steeper or changes more quickly in slope. 17. Level soil and land within bunded plots and fields so water depths are uniform so less water is needed overall. 18. Add smaller bunded rice plots/paddies to large fields to improve water control and depth. 19. Build strong, clear soil bunds with defined cuts and channels to manage water flows between bunded plots. 20. Add walkways through fields to enable visibility of where water is flowing (necessary in 3-metre-high sugarcane). 21. Cycle water between areas (applying wet/dry cycles which can raise rice yields and save water). 22. Schedule dry season non-rice crop irrigation (deficit irrigation). 23. Attempt surge irrigation where initial surface wetting then subsequently seals soil surface to reduce deep seepage. *Irrigation canals and gates* 24. Respond to rainfall events (close intakes to allow irrigators to reduce inflows during floods). 25. Add more canals to improve location of water to different parts of the irrigated area. 26. Add, resize, or improve canal gates to add adjustment control of water, sized to match ratio of flow-to-area. 27. Maintain and clean canals and gates to improve control and switching of flows. 28. Line canals if condition deemed to be significantly affecting seepage and control – not as a reflex action. 29. Add extra freeboard to canals where irrigators are throttling back to reduce over-topping. 30. Add control structures or multiple inlets (spiles, syphons) to fields to improve uniformity of flow rates along field edge.

Note: These reduce net demand via substitution (different varieties), dematerialisation (deficit irrigation) and retrenchment (reducing areas irrigated).

Source: From research conducted in Tanzania under the RIPARWIN project (McCartney et al., 2007).

Third is the control or regulation of withdrawal of a natural resource from its common pool. Without this, or with a weak form of regulatory control in place, higher consumption within the resource system is potentially enabled. In other words, an improved efficiency (via actions in Table 18) can reduce consumption if withdrawal is throttled downwards over time. In an irrigation system, this would manifest itself as a set of headworks on the river (known as an offtake or intake) that is sized appropriately and can be adjusted downwards temporarily or permanently as efficiency within the system begins to improve. This type of infrastructural change might be accompanied by changes in water rights.

The fourth dimension is the technical ability (or inability) to precisely control the flows, switches and locations of resources passing through a large technological social-ecological system. Control of a resource such as water is addressed through quite a considerable literature going back two decades (e.g. Plusquellec et al., 1994), from which two points are made: first is the distinction given to role of assemblages of technology – a topic picked up below. The second point is that technologies and activities to adjust the switching of fractions do not predictably give the precise outcomes we seek (yet might give some sense of what contributes to overall resource control). This is for two reasons which, for ease of classification, might be termed non-human and human (or hardware/software). In the non-human instance, an excessive number of natural and artificial factors are at work within large nested systems. For example, land levelling of irrigated fields improves the uniformity of infield water infiltration and contributes *generally* to a reduction in non-beneficial consumption (by reducing dry and flooded spots inside the field) but does not 'buy' an exact gain in water control and efficiency alongside other factors that might not be quite right such as wet soils following rainfall likely to lead to water running off fields into ditches. This also means that it is exceptionally difficult to control switches between withdrawal fractions (Section 4.7.2). Pease (2012) also makes this point.

Fifth, I argue in Lankford (2006, 2012) that fractions are coupled. This means they move in lock-step with each other but in uncertain ways. For example, reducing the beneficial consumption of crop transpiration can reduce non-beneficial consumption. Expressed differently, a reduction in the command area, or deficit irrigation would also reduce the recoverable fraction.

Sixth, control of the placement, timing and quantity of other inputs alongside the main resource is required to produce a crop (or sequester carbon). Without this, the productivity of both the main resource and other inputs is constrained by poor co-ordination. As an example from fertiliser and irrigation illustrates, atmospheric losses of nitrogen increase if urea is not followed soon after by an irrigation to draw the fertiliser into the soil matrix.

Seventh, control of water can also be found in other types systems, urban and environmental. Pittock and Lankford (2010) explain how water demand and share management can be applied to sustain or increase environmental goods and services – all predicated on a more judicious placement of water through time and space.

4.6.6 Lacking data: accounting, measuring and monitoring

The uncertainty at the heart of resource efficiency complexity and the paracommons is, to a great extent, derived from the relative failure to embed appropriate schemes for measuring and monitoring the intricacies of resource use and sharing. van der Kooij *et al.* (2013) itemise the lax methods taken by scientists in assessing water use and savings, concluding that results are not generalizable or trustworthy.[10] However, their work applies to studies of efficiency. In most irrigation systems in the world, efficiency and productivity go unmeasured because systems (inflows, areas, evaporation, drainage etc.) are not monitored sufficiently.

This task should not be underestimated. The fugitive nature of water and carbon moving through society, landscapes, atmosphere, soils and geology in different quantities and forms means the flows and fractions of resource withdrawals are unexpectedly difficult to trace. An incomplete response to this challenge undermines the reliability of efficiency and productivity computations in addressing natural resource management and sustainability. The answer as, Skaggs *et al.* (2011) cogently argue on precisely this issue is to determine the actual not the modelled use of resources. Therefore the following eight paragraphs cast the accounting, measuring and monitoring of resource performance and efficiency as aspects of 'effort'.

First there is the question of how accounting theory should be designed for scalar-related and cumulative inflows and outcomes alongside against the details of tracing intermediate water flows and losses at different stages. For example, the IWMI-WA method outlined in Section 4.7.2 appears to be designed primarily for final dispositions the river basin scale – and that it cannot capture the details of water savings, economisation and recovery within nested scales. This is not to say that the IWMI-WA method over-simplifies the basin scale – on the contrary, see Karimov *et al.* (2012) for the work requirement to determine what is taking place in one case study, the Fergana Valley.

Second, in terms of accounting theory applied to one sector or type of system, an intellectual effort is required to discern which model (if there is more than one) is required for a given situation. Irrigation exemplifies because efficiency theory uses both 'classical efficiency' and 'effective efficiency' computations to respectively include and exclude recoverable losses in the denominator. This difference means that classical efficiency values are lower than effective calculations, and that the former can be used for evaluating irrigation schemes (Lankford, 2012b) while the latter is utilised for assessing the impact of irrigation on river basin water balances (Keller *et al.*, 1996; Haie and Keller, 2008). The emerging concern, however, is that water users, engineers and other stakeholders are not actively engaging with these choices and employing them appropriately (Molden *et al.*, 2010).

Third, once the accounting frame or theory has been selected (though in irrigation this is the subject of an intense debate; see Gleick *et al.* (2011), it does not then follow that methods to measure the quantities and dimensions of the different fractions are agreed by all parties or easily determined. For example, I believe that an erroneous emphasis on sampling losses in irrigation canals fails to record the myriad and micro ways that water is reused over a season within an irrigation system (Lankford, 2012b).

Fourth, it is possible to argue that ongoing day-to-day management of systems suffers from a lack of functioning durable hardware and software for monitoring resource use. While hydrologists lament the lack of river-flow monitoring (Hannah *et al.*, 2011), this supply-side gap is minor compared to the lack of equipment for monitoring the demand-side and sharing of water and its inefficient tare fractions. Ryan *et al.* (2010) make similar comparisons to the problematic monitoring of carbon in forest stocks and 'the difficulty and expense of tracking forest carbon, the cyclical nature of forest growth and regrowth' (p.9).

Fifth, leading on from the previous point are questions of the appropriate infrastructure for measuring flows, goods, services and losses. This is not simply a matter of installing measuring structures at a sufficient density – rather it relates to the variables required, the robustness and ownership of infrastructure, the extent to which data collection infrastructure fits both together and the system's characteristics, and the extent to which proxy measures might be used (e.g. rate of planting and watering for the rate of losses, see Section 7.3).

Sixth, even if collected, quantitative measures must be treated with caution. Because context is so variable, one might argue that absolute measures of efficiency are difficult to interpret and hold little meaning. On the other hand, by being context specific and 'relational' (Geels, 2010), a particular system's efficiency, compared to itself (in a trajectory of attempts to improve systems) or to other similar systems, relative measures might be more useful. For example, irrigation efficiency as an absolute measure (e.g. 60 per cent) is arguably meaningless. Instead a benchmark against previous or to neighbouring conditions (e.g. 79 per cent of best in the region) might assist resource users and their service providers (Lankford, 2006; Solomon and Burt, 1999).

Seventh, measuring losses may not be the same as measuring who gets the material gain of a performance gain. Detecting who ended up with different sources of paragain builds upon appropriately tailored accounting and measuring methods, as discussed in Section 7.3.

Finally, with regards to the reform of institutional arrangements of common pool resources, an 'effectiveness' gap exists. This may be summarised in the question – do new technologies, devolved institutions or markets result in water being used more efficiently, equitably and productively when accurately quantified? This gap appears throughout CPR research and implementation, and was alluded to by Dolšak *et al.* (2003), in their concluding section on 'developing new methods'. In my review (Lankford 2008) of the Warner 2007 book on multi-stakeholder platforms in water, I was clear of differences between CPR principles (Ostrom, 1990; McCay, 1996) and of the measurement of their outcomes on material resource patterns and productivities. I wrote this review:

although the book attempts to answer the question of whether multi-stakeholder platforms (MSPs) make a difference, this is insufficiently traced in terms of water productivity and managerial performance outcomes (but it should be readily admitted this is very difficult to achieve) in the various case materials. In other words, did the MSP result in enhanced river basin performance no matter how the latter might be measured?

Because resource accounting and measurement is so central to understanding the paracommons, and its distinction from the commons, the issue is taken up again in Section 7.3 on efficiency kinetics.

4.6.7 Human expectations of control technologies

Emerging from the previous two sections covering resource control and metrics, is the risk of unmet expectations surrounding the changing of technologies of water control that intend to produce more and consume less. Briefly, I relate the likelihood of thwarted expectations to two scalar effects related to technologies of resource and wastage control.

The first arises from the failure to distinguish between technologies that work at the micro-scale (household or irrigator's field) from assemblages of technologies that work at a higher system scale (city or whole irrigation scheme). In understanding this flaw, the technological rationale for improving efficiency needs to be carefully explored – particularly whether actions to raise efficiency are indirect or purposive. I take as axiomatic that resource users, service providers (e.g. engineers) and policy-makers regularly alter their practices to indirectly affect the behaviour of whole systems. This is because humans tend towards ideas and practices that might save labour or produce more from less. But with respect to the uncertainty of the paracommons, I take this approach as largely 'indirect' and not particularly strategic because of the problem of orchestrating interventions in the face of spatial and temporal scales. Thus, technologies may be simple and targetable for a smallholder's field with rainfed maize, no fertiliser and a hand-hoe to till with. However, scaled up to a large 3000 hectare irrigation system sitting within a 1000 square kilometre catchment, and the nature and size of expectations of what is, and what could be, grows substantially.

Here, the insight for REC and the paracommons is that while at the micro-scale (one field or one house), users and their support agencies intend to 'do better' on a daily or weekly basis assisted by relatively clear feedback signals at the unit scale, this intention does not translate to the larger scale. At larger environmental, system and societal scales over longer periods, emergent and unpredictable feedbacks and fudging of boundaries result in disagreements and unforeseen results. The reworking of systems that are invariably large scale, neighbourly, coalescing and nested implies that, in the absence of thoroughly comprehensive technologies of control and measurement, the raising of efficiencies of whole systems and sectors is characterised by uncertainty and unpredictability, with the likelihood of expectations dashed or methodical experimentation undermined.

The second scalar dimension of technology relates to the cost-effectiveness of pursuing solutions at the lower scale; in other words, total losses arise via a cumulative pyramid of smaller losses at the 'bottom of the system'. A good example of this can be seen in how water boards in urban environments are quick to repair mains water leaks (being highly visible and easy to locate) but are more ambivalent towards many small leaks at the household and street level. The same tension is found in irrigation: engineers are quick to point out the need for membrane lining of the single main canal and tend to 'ignore' myriad in-field features (ridge and furrows, bunds, gradients, land-levelling) that lead to poor infield water control. The significance of this scalar cost-effectiveness lies in the unknowns and attribution of both the costs and effectiveness of ways forward – particularly when both cost and effectiveness have to be judged in terms of who pays and gains at different nested scales – the farm, irrigation system or river basin. Human operational and maintenance factors also affect water control. Even if design is near-perfect, it would be inconceivable to expect thousands of irrigators to operate their fields, furrows and ditches with military exactness. The literature of irrigation practice, certainly during the latter half of the twentieth century expended much effort in casting 'O&M' as an issue in irrigation performance and productivity (examples include Sagardoy and Bottrall, 1982; Biswas, 1990).

This mismatch between expectations and extant outcomes are often expressed in different types of policy and training packages that in turn seek efficiency improvements by over-simplifying interventions or by adopting step-change technologies and approaches. I briefly describe three examples:

- Water control is intended to be raised by adopting wholly new systems that utilise different field level control technologies – for example, switches from flood (canal) to sprinkler or micro-irrigation (drip). Donors (e.g. JICA and IFC) seem particularly keen on drip solutions to large scale water efficiencies.
- Water control at the sectoral or system level is managed when decisions are made about which particular projects (if more than one) receive more attention than others. This type of control is central to the paracommons when offsetting is part of a programme to reduce aggregate losses or to raise production. An example might be when a REDD+ carbon project is closed down in favour of a new one initiated.
- Based on the author's experience at University level in studying and teaching irrigation, and as an external examiner, there is commonly insufficient space in the irrigation curriculum to address the multiple dimensions of efficiency. It is pedagogically more expedient to package efficiency in distinct narratives for example, classical versus effective irrigation efficiency or drip versus canal. However, binary views of efficiency do not allow for much interpretation of complexity.

On the other hand, deciphering technologies of production is key to identifying what underpins current practices, how likely an increment in productivity and efficiency is, and how studies might guide improvements. Trawick (2001) cogently describes how irrigation in the Andes presents a variety of physical and social methods of controlling water, where the gradients and plot-sizes in the landscape allow irrigators to pass water to each other in relatively informal ways (spillages, streams, groundwater) that some 'pipe, canal and drain' irrigation engineers might not recognise. Intimately connected to an ecology of water control are the knowledges and social practices of water users that dictate, shape and mediate how a resource is managed, and more particularly how a programme to raise efficiency of resource use is adopted and adapted.

Interpreting current water management accepts path dependency (Heinmiller, 2009) in incrementally navigating to new ways of managing water rather than to radically depart from current procedures thereby risking unsustainable rapid change. However, donor interest in irrigation efficiency improvements seems quite immune to this observation, favouring radical modernisation in the belief it will achieve 'efficient irrigation'. While wholesale technological change might work for the private sector, one must question the extent to which it is a solution for public irrigation managed collectively by smallholders.

4.7 Quaternary: resource expression

The quaternary group of REC factors covers three types of resource expression which depend on the type of resource and types of losses. For the purpose of the paracommons, a resource expression conveys the manner in which a resource 'resolves' into outcomes during a conversion of a principal resource (such as coal or groundwater) into the main goods/benefits and its wastes and wastages. These three expressions are addressed in sub-sections below.

4.7.1 Three-way resource conversions

The 'simplest' type of resource expression is where the resource is directly depleted in its conversion with only non-recoverable wastage and recoverable but possibly unwanted waste comprising the 'loss'. An example is coal; the 'good' is heat and energy, the wastage is gas (e.g. carbon dioxide) and wastes comprise ash and cffluents. One might argue that these three-way resource conversions (to goods, wastage and waste) are not ideal representations of the paracommons because of the relatively simple resolutions and dispositions of goods and losses. However, although broadly 'three-way', with further innovation such conversion systems can become more complex. With this in mind the three efficiency goals that add complexity are: a) the conversion of more of the solid coal to useful heat and energy; b) the transfer of wastes to other industrial processes that can make use of them (industrial ecology), and c) the treatment of previously unrecovered wastages (for example, carbon dioxide capture). Further discussion of this type of conversion-and-loss is found in Sections 5.2.2 and 5.3.3.

4.7.2 Fractions as portions and pathways: tracing multiple conversions

The second type of resource expression comprises quantitative fractions of a resource (the other, services, is addressed next). Fractions (Willardson, 1994; Perry, 2007) describe portions of a common pool resource resolving to different outcomes (or 'dispositions', Frederiksen and Allen, 2011) as a result of a withdrawal and conversion-and-loss process. Exemplified by irrigation, water in a river or aquifer is converted into outcomes such as agricultural products via beneficial consumption (evapotranspiration) and various types of wastes (e.g. a recoverable flow such as found in a drainage ditch) and wastages (e.g. non-beneficial evaporation) of the original resource. For the purposes of this section, I introduce two accounting methods in the literature termed here as the Perry and IWMI WA (water accounting) approaches, discussed below.[11]

Water accounting (e.g. Perry, 2011; Molden and Sakthivadivel, 2006; Foster and Perry, 2010) see fractions as portions of water withdrawn ending up at different mutually exclusive dispositions (e.g. beneficially, non-beneficially, within the irrigation system or within the basin, and so on). Table 19 gives the Perry (2007) taxonomy for physical portions defined by where water 'processually' ends up and whether it is ready or not to fulfil further work. Table 19 defines two main types of process outcomes, consumed and non-consumed, which subdivide into other outcomes. The consumed fraction includes the beneficial and non-beneficial outcomes (to which I have taken the liberty of adding process and non-process beneficial consumption from the IWMI approach – see also Molden and Sakthivadivel, 2006). The non-consumed fraction is either recovered/able or not recovered/able. The classification also implies a physical and geographic disposition; examples given by Perry include water moving to the sea, to deep aquifers or to saline aquifers.

However, I have labelled this section 'Fractions as portions and pathways'.[12] This is because pathways are used by the paracommons framework to explore complex options associated with switching between and altering fractions. Casting fractions as both portions and pathways allows eight points to be made (discussed later in this section). More particularly, the idea of fractions as pathways is central to the concepts of liminality and the paracommons. By contrast, it is the emphasis on (fewer) portions in the Perry/IWMI water accounting frameworks that distinguish the latter from the paracommons approach.

The IWMI water accounting framework is shown as a generic finger-diagram in Figure 35. It is broadly similar to the Perry/UNSEEAW framework in that depletion is divided into three portions (beneficial process, beneficial non-process and non-beneficial depletion). By combining these three into water that is 'depleted', it then leaves three remaining portions; the 'outflow' (comprising

Table 19 Water accounting for water withdrawals

Consumed fraction	*Beneficial consumption:* this can be divided into either 'process' or 'non-process'. 'Process' is water evaporated or transpired for the intended purpose – for example, evaporation from a cooling tower, transpiration from an irrigated crop. 'Non-process' beneficial consumption is all water not intended for a key objective but nevertheless is associated with benefits – examples include evaporation from reservoirs or riparian vegetation.
	Non-beneficial consumption: water evaporated or transpired for purposes other than the intended use with little or no benefit. Examples include evapotranspiration from weeds and wet soil.
Non-consumed fraction	*Recoverable fraction:* water that can be captured and re-used – for example, flows to drains that return to the river system and percolation from irrigated fields to aquifers; return flows from sewage systems.
	Non-recoverable fraction: water that is lost to further use – for example, flows to saline groundwater sinks, deep aquifers that are not economically exploitable, or flows to the sea.

From Perry (2007), Perry in Frederiksen *et al.* (2012) and Karimi *et al.* (2012b)

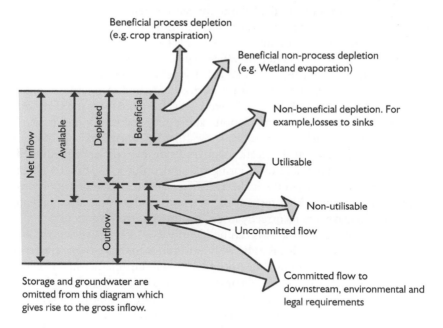

Figure 35 Finger diagram of IWMI water accounting

'utilisable and non-utilisable'), non-committed flows and committed flows. An example of a finger diagram and the latest thinking on the IWMI framework can be seen in Karimi et al. (2012b). For the purpose of this book, the key difference between Perry and IWMI frameworks is that the latter focuses on final depletion at the basin scale while Perry allows for recovered flows, explained here:

> This is different from other approaches that account for water withdrawals. The advantage of using depletion over withdrawals is that it is hard to estimate return flows and recycling of water within the system. By focussing on flows across domain boundaries, withdrawals and recycling are internalised within the IWMI WA.
>
> (Karimi et al., 2012b, pp.77–78)

'Fractions as both portions and pathways' play eight important roles in resource efficiency complexity and the paracommons. First, with regards to 'portions' there is no doubt that 'fractions' as posed by Perry reveal the complexity in conceptualising and calculating water flows and efficiency. Fractions define outcomes for water from a basin perspective (e.g. beneficially and non-beneficially consumed). Although not overtly employed to calculate efficiency, fractions demonstrate that efficiency cannot simply be divided into the ratio of beneficial fraction divided by the withdrawn fraction. Fractions show there are other possibilities for determining the constitution of the denominator – which leads to the choice behind the two paradigms of classical and effective efficiency (Lankford, 2012b). (See Figure 15 for the computation of these two ratios). By defining fractions, resource flows are then subject to debates about water performance and accounting at different scales (e.g. Perry, 2011; Molden and Sakthivadivel, 2006; Foster and Perry, 2010). Fractions also reveal that computations and policy matter for the right scale. The fractions approach, for example, shows that policy-makers supporting drip irrigation presuppose canal irrigation suffers from low efficiency predicated on a classical computation and by ignoring the return flows associated with canal/gravity irrigation, may overlook the risk of greater resource depletion despite technological change. Not recognising fractions and associated computations allows policy-makers to incorrectly insert inappropriate interventions.

Second, as portions, fractions can be identified as already existing within the common pool resource that is to be withdrawn prior to the resource conversion-and-loss. Thus in irrigation, the specification of the 'inefficient' or extrinsic fraction is integral to the design of irrigation systems; this volume has to be located within the common pool before defining the aggregate (gross) withdrawal volume. This allows an irrigation scheme to legitimately claim a fraction to meet putative within-system losses. In turn, if these losses are forestalled, reduced and released (or incorrectly overestimated in the first place), savings are then the subject of an ownership debate and imply decisions regarding when a common pool resource begins to be owned in the chain of harvest and conversion, an issue wrestled with in the Montana vs. Wyoming case (and addressed in Chapter 6).

Third, fractions as portions are connected to the question of accommodating scale and nestedness, neighbourliness, described earlier. The significance of a fraction arises because it 'flows' (or routes) to the next scale up. For example, a recoverable fraction is lost from one farmer's field but gained by the river catchment. Furthermore, to continue the point in the previous paragraph, the inefficient fraction to be withdrawn by an irrigation system means that this can be locked into the water right for that system, which paradoxically reduces the access to water of users *upstream* of the system of interest. The issue here is also about the scale at which water accounting is conducted – both when analysed at a whole basin level but also when analysed at smaller unit level and then recombined synergistically at the basin level. UNEP (2012, p.77) concluded on these problems of aggregation and disaggregation both spatially and temporally.

Fourth, fractions as portions 'flow' from a common pool to different dispositions and therefore take some kind of physical pathway through a system; however, they do not necessarily flow in separated quanta and as easily identifiable 'flows'. (This observation is the beginning of the idea of fractions as pathways addressed in the next three paragraphs.) It is impossible to go to part of an irrigation scheme (e.g. the main canal) and 'see' or measure the different fractions. A water accountant and irrigation manager (or farmer) have their own perspectives on water travelling through an irrigation scheme and it is this difference between them that opens up complexity and pathways as options. A few examples of the many ways a farmer sees water and water losses include seepage, weed growth in canals, dry spots in the field, early cessation of irrigation, delayed timing of delivery and so on. As will be explained, it is very difficult to translate these irrigation phenomena into accounting fractions.

Thus, when fractions are seen as pathways (in other words as options), greater complexity is revealed. This leads to the fifth point regarding how fractions as portions are managed (i.e. one is reduced and another increased) and it is here that we find multiple options alluded to in the final three sentences of the previous paragraph – and behind them many other choices over technology (e.g. land-levelling, longer furrows, fitting land preparation and irrigation around rainfall events and so on). It seems that water accounting envisaged by these and other authors is unable to comment on how flows of resources, passing through the nested scales of fields, sub-systems, systems and basins, relate to each other. The IWMI WA method is explicit in stating it cannot 'report on irrigation efficiency' (Karimi *et al.*, 2012b, p.77).

Sixth, these options must be seen in the light of expectations to raise efficiency, be more productive and reduce aggregate consumption. This implies a transition between a current situation (where efficiency is believed to be improvable but most likely not measured or fully understood) and a future scenario (where efficiency is intended to be higher, but most likely not measured or where aggregate consumption is also not controlled for). This transition between one scenario and another, driven by expectations and assumptions (or 'the political economy of promise' (Leach *et al.*, 2012)) is central to the idea of liminality in the paracommons. It is why Tables 4 and 13 employ a starting point and future to calculate paragains and why Table 16 questions which start and end point should be employed in calculating efficiency changes and resulting paragains.

Seventh, the IWMI WA and Perry methods despite accounting for 'outcomes', do not comment on the destination objective of the 'salvaged' resource, how this is achieved (except by reducing non-beneficial consumption, which is only one option of several) and what parties are involved. Neither framework is able to comment on how the finer details of water management influence what happens to water made available by efficiency changes (the paragain).

Finally, it is worth distinguishing between this section discussing 'fractions', and the next section on 'services'. In this section exemplified by irrigation, flows of resources remain principally as water (albeit delayed, salinized or polluted). The only different form is the vapour 'wastage' arising from evaporation and transpiration. In the Section 4.7.3, services are the emphasis. The best example of this is carbon which can manifest itself as multiple forms with different benefits and functions for a variety of interest groups: for example, as soil fertility via organic matter content; as timber via forest products, or as avoided deforestation valued by carbon markets. With regards to carbon sequestration pathways, an example may be found in Law and Harmon (2011).

Therefore, 'fractions as portions' do not sufficiently explore the extent of choice present within complex social-ecological systems or the reciprocal relationships between irrigation systems and basin dynamics. Water ending up as different 'outcomes' should be seen as different from the potential pathways, options and destinations available. This variety of possible pathways for resources to resolve themselves into generates the 'liminal space of potentials', a defining feature of REC and the paracommons. Taking water in irrigation as an example, it is possible to discern 16 different pathways that resources, wastes and wastages may fall to (in Section 5.2.3 and Figure 38). Furthermore, by going beyond fractions as portions, we can see, for example, that a reduction in aggregate depletion can be arrived at through several means. Figure 36 figuratively shows that a transitive field or space consisting of a current scenario and a future scenario sits between 'inflows' and 'dispositions'. This liminal space contains multiple pathways and strategies for managing resources in attempting to raise performance while controlling for consumption.

4.7.3 Ecosystem services as pathways and options

Briefly introduced in Chapter 3, efficiency and productivity complexity also arises when resource conversions generate ecosystem services of qualitatively different kinds (MEA, 2005). The fixing of carbon dioxide by plants provides examples such as biodiversity, timber; food production, short and long-term sequestered carbon and soil-carbon mediated changes to watershed streamflow. These occur during different stages as a result of carbon fixation, metabolism and recycling. Efficiency and productivity interests play a part when we seek

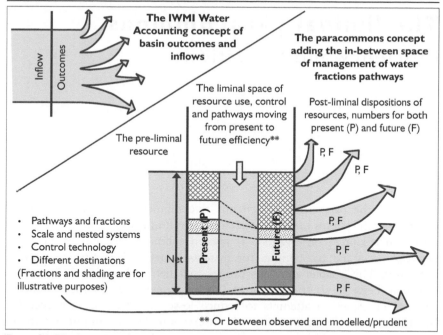

Figure 36 The liminal space in between water inflow and outcomes

to boost one ecosystem service over another – in other words, when carbon sequestration takes priority over timber production, and when an increase in carbon sequestration is sought in the future over today's rates of sequestration. Scholars have studied how land use change and other management decisions have influenced trade-offs between ecosystem services (Gordon *et al.*, 2010). However, it is because ecosystem service paragains (salvaged losses) are different from physical paragains that the paracommons discussion about ecosystem services is collated under one section in the next chapter to which the reader is referred.

4.8 Chapter review

Twenty sources of complexity arise during, and mediate, attempts to understand and boost the efficiency of natural resource systems and sectors. These factors are classified into four types: definitions, language and understanding of efficiency; the nature, context and purposes of the resource efficiency; scale, nestedness, neighbourliness and time; and resource expression. These factors individually and collectively define the five features of the paracommons; high complexity; systems in transition (liminality); competition over paragains created by efficiency gains; inter-relatedness of users through losses and the reduction of losses; and the paradoxical thwarting of policies to raise efficiency. The next chapter explores the second of these ideas; the liminal nature of the paracommons.

Chapter 5

The liminal paracommons
Efficiency and transitions

Recapping, the paracommons contains five themes: 1) resource efficiency complexity; 2) systems in efficiency transitions (liminality); 3) implications for CPR and resource conservation; 4) revised subtractability, interconnectivity between users and competition for material gains, and 5) efficiency policy paradoxes. This chapter develops the second theme, the idea of systems in transition between a current state of (lower) efficiency and a future state of (higher) efficiency in attempts to reduce losses and free up resources for further use, including common pool conservation. This feature of systems in transition is captured by the idea of liminality (next) and three types of liminal spaces (Sections 5.2). Section 5.3 probes resource systems by asking how they distribute resource paragains to one or more of the four destinations.

5.1 Liminality as complexity

Returning to Figures 1 to 4 and Figure 11, I propose that the interjection of resource efficiency complexity (REC) for systems undergoing efficiency transitions establishes and defines the paracommons. Recognising 'transition', I collate ten substantive reasons for choosing liminality as a key feature of the paracommons. The first four are the most important from which others emanate:

1   Physical transition and mathematical. Threshold and transition are mathematically embedded in the centre of the efficiency ratio as the vinculum. The vinculum is the threshold between the denominator and numerator connected via a conversion-and-loss process. Physically, resources move from their aggregate gross form (denominator) in the common pool to their net, intrinsic beneficial form (numerator) in the produced good or service. Resources cascade and transit through complex switching and converting systems (e.g. irrigation schemes and forests) to final dispositions. In this transition an extrinsic, rejected tare (or loss, waste and wastage) fraction is generated which is in the denominator but not in the numerator. While the efficiency ratio contains information about the performance of this process, the ratio also implies a loss fraction is

potentially salvageable as a material gain for further use and competition –
a detail not well recorded in the efficiency ratio.

2   Temporal – progressive and comparative. The raising of efficiency is
implicitly part of the managerial promise to transit from a current state
of (in)efficiency to a future higher efficiency. The efficiency ratio implies
that a goal exists; the ultimate being a perfect conversion of 100 per cent
of the common pool used into beneficial goods or services. (Even without
this perfect goal, through pervasive environmental narratives and lack of
measurement we can be dissatisfied with current performance.) Thus
science and society is interested in an improvement in efficiency over time –
reworking a system from a current to a future performance. This denotes a
temporal threshold and therefore transition between two states: the current
or earlier state of 'being inefficient' and the future state of 'being efficient'.
This performance gain introduces the problem of how science and society
judges systems politically and culturally – and how this compares with views
held by local resource users and NGOs. The political and policy intentions
underpinning this progression are discussed in points 9 and 10.

3   Temporal – transitive and transient material gains. In natural resource
systems, the efficiency increase of the previous point is paralleled by a
material gain or 'tare' revelation – the 'freeing up' (even if immediately used
again) of a 'waste/wastage' fraction. However because of the complexity
factors highlighted the previous REC chapter, questions abound about the
transitive and transient nature of losses in limbo moving to new patterns
and uses within that change programme. Examples of these liminal
questions are; how do the loss fractions become potentially salvageable
and then salvaged; how transient or ephemeral is the salvaged resource;
what technologies control and direct the losses; and who is their legal
owner. One might interpret the salvaged losses (paragain) to be a fleeting
liminal resource; identifiable within the main canal flow and once it has
decomposed into a drain yet difficult to trace, store and redirect while
transiting these two states.

4   Shifting gain destinations. Material gains from raising efficiency migrate
between, or to all, four destinations of efficiency (proprietor, neighbours,
common pool and the wider economy) depending on resource properties,
scale and scarcity – but also on our interpretation of, and response to,
those circumstances. Thus, one REDD+ or irrigation project might send
paragains to a different destination than another. This unpredictability of
where the gain ends up establishes consumptive paradoxes As introduced by
the Jevons Paradox, a liminal space exists between our intuitive expectations
of conserving the natural commons against rather disappointing increases in
aggregate depletion. Along with the idea of unforeseen gain distributions, I
argue that these paradoxes signal any number of outcomes that befall large
social-ecological systems undergoing efficiency improvements (hence the
idea of a multiplicative liminality).

5   Neighbourly uncertainty. In Section 4.6.3, the idea of neighbourly distribution was introduced – that reworking of a proprietor's efficiency disproportionally affects the distribution of resources for users in that system and for users in closely and less closely connected neighbouring systems. It does this in ways that are difficult to predict. The apple example at the beginning of the book shows how different options set up uncertainties for the different actors depending on their immediate or less immediate relationship with the householder.

6   Normal assumptions about waste and wastage. An uneven transition takes place amongst different stakeholders around the importance of waste and wastages partly mediated by scarcity, technological innovation, costs of salvaging losses, and professional training. Thus some irrigation engineers believe that drainage water is unfit for beneficial use and should return to the rivers while farmers readily take the opposite view – drainage water is a valuable resource. Both sets of beliefs are moving through time, however, and both change and learn from each other.

7   Control and data impreciseness. In Sections 4.6.3 to 4.6.6, I argued that selecting resource control interventions do not purchase exact gains in efficiency or precisely steer resources and their loss fractions to intended dispositions. Efficiency fractions are coupled with each other and their precise place and form over space and time are exceptionally difficult to ascertain, steer and control. Because of this, there are disconnects between what appears to be going on, what are available as interventions and what ultimately transpires.

8   Switching services. In ecosystems, different services can be emphasised and deemphasised creating, often unwittingly, less important services and subsequent shifts between numerators and dominators.

9   Efficiency policy prefiguration. There is a gap or space between a policy intention and the final policy-driven outcome made real by society's and science's rather dubious track record in fully accounting, measuring, managing and controlling resources and their fractions while accommodating spatial nestedness and neighbourliness, time and boundaries. This topic is further addressed in Section 7.1.1.

10 The politics of efficiency. While it might be possible to charitably defend policy as being typically 'utopianistic', 'behind the curve' or unappreciative of the complexity of resource efficiency (the previous point) there is another more critical interpretation of policy addressing resource use efficiency. In this situation, the 'betwixt and between' of the liminal paracommons offers political players (such as donors or policy-makers) opportunities to insert themselves into the politics of resource efficiency, even maladaptively. This topic is picked up in Section 7.1.2.

In work on 'stages' within liminality, applied to an anthropological analysis, van Gennep (1909) describes ideas that seem to fit very well the paracommons.

A common pool resource sits as a pre-liminal whole, then is subjected to harvest and then 'separates' into many possible streams (with many avoided, forestalled, productive, recovery and consumptive possibilities) and then re-assimilates post-liminally into fewer environmental outcomes or dispositions; depletion, the common pool or the product.

Taking the earlier example of the apple and apple core, one observes a physical transition of the apple core from one location to another depending on what happens within the householder. So while there is always one apple core critically it is a) the uncertain futures and b) the movement of the core through the proprietor system (householder) to other potential destinations that determine liminality. With regards to uncertain futures one compares the reactions of the city dump recycler between her witnessing simply a longer delay ('the discarded apple will now take four hours instead of two hours to make it to the garbage dump') with a new waste policy ('all householders are to throw away less food or to recycle more food waste at home'). The latter is accompanied by much higher levels of uncertainty. The insertion of an efficiency drive at the household gives the city dump recycler greater concern regarding what is about or not about to transpire. While the apple core remains physically one thing, it 'turns up' in two places; the proprietor and the city dump (or one of the parasystem destinations). Thus there are now at least two expectations for the apple core; the housekeeper responding to new instructions about waste and the waste picker concerned that less food will arrive at the city dump. Prior to the efficiency drive, the unvalued apple core formed only one expectation (or no part of any expectation) as it simply wended its way from the house to the dump as it had many times previously. There is also a proprietor expectation regarding the apple core; the householder naturally expects to have and consume an apple comprising flesh and core. The proprietor's expectation arises because the apple core forms a commonplace part of the surrounding flesh – it supports the apple flesh (without specialist vacuum wrapping, apple flesh without a core would rot more quickly) and secondly for some (more efficient) apple eaters; the core is edible – it is the apple). Crucially, the apple core and apple are not separated from each other prior to entering the proprietor system and then discretely sent to each destination. On the contrary, the fate of the apple core is mediated by the proprietor and any matters that influence the proprietor.

This dimension of 'expectations' surrounding multiple possible realities arising from resources transiting systems undergoing efficiency changes is the salient feature of the paracommons. The transiting apple core, potentially shifting in space, use and amount, is the liminal object at work here – it is part of and creates these competing expectations held by users in different destinations. So while there is only one apple core, it arises in the minds of all four users, until it is completely disposed of (composted, incinerated or eaten by householder or dump picker). Between arriving at the proprietor's house and final disposition, the 'to-be-wasted/to-be-salvaged' apple core is, from the combined perspective of the four destinations/users, betwixt and between. The potential parallel fates of the apple core create this particular liminal paracommons.

5.2 Three types of efficiency paracommons

5.2.1 Introduction

The 20 building blocks of resource efficiency complexity variously combine to produce three types of paracommons each defined by efficiency 'liminal spaces' (productive-consumptive, multi-path and multiservice). Within REC and the paracommons, a liminal space is a transitive threshold through which resources and waste/wastage fractions pass to an array of productive and environmental dispositions depending on the nature of an efficiency improvement effort in a given context. A liminal space signals difficulty in accurately predicting an exact set of outcomes including verifiable changes in system efficiency and common pool consumption. This difficulty occurs because the harvesting and management of natural resources should be seen as taking place within neighbouring and nested, recycling social-ecological systems. Thus while the efficiency of a single irrigated field can be accurately manoeuvred and contains less 'room' for a great number of unresolved potentials, a larger irrigation system within a larger catchment is much more complex. This complexity leads to much greater latitude for different outcomes to unpredictably transpire. These outcomes pertain to both users within the irrigation system, users immediately dependent on downstream recovery and users elsewhere in the river basin. It is this growing uncertainty and number of different potential outcomes that builds the thickness of the threshold or liminality between what is expected and what eventually transpires at different locations and scales.

Referring back to Figure 19, all three liminal spaces are united by the first three groups of 17 sources; the primary group of language and knowledge, the secondary group or resource properties and context and the third group or accommodating scale and boundaries. Although the three liminal spaces are similar with respect to uncertain outcomes, there are differences. The first liminal space describes difficult-to-predict and paradoxical consequences for aggregate consumption and production. This is largely the result of dynamic rebound where the resource and conversion efficiency feed each other in unforeseen ways. The second liminal space brings in the influence of 'fractions as pathways' within a given abstraction-and-conversion system (e.g. irrigation scheme). The third space draws on society's views of the ecosystem services that a resource might imbue a region, zone or landscape.

5.2.2 Productive-consumptive paracommons

The 'productive-consumptive paracommons' arises from the three-way conversion options that some resources exhibit when used and consumed (for example coal burnt in a power generating station). Unlike the multipath commons (in the next section) there are three main pathways within the conversion process as shown in Figure 37: [1] the unrecoverable wastage pathway; [2] the 'goods' production pathway, and [3] the recoverable wastes

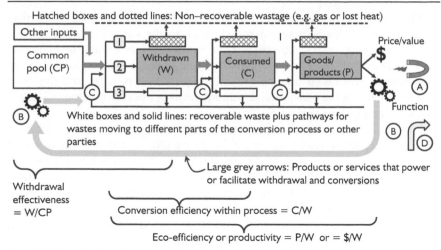

Four efficiency-induced consumption paradoxes

A: Efficiency-induced consumption of outputs from lower prices and more products

B: Efficiency-induced withdrawal and use of inputs from facilitative functions (e.g. energy)

C: Pathways of wastes recovered then converted to products

D: Offset-efficiency is claimed, yet errors might induce higher total consumption

Figure 37 The three pathways of the productive-consumptive paracommons

pathway. The term 'productive-consumptive' indicates that attempts to boost production by raising efficiency have a propensity to raise resource withdrawal and consumption, explained below.

Relevant to resource efficiency complexity, there are four paradoxical properties or processes that explain consumptive consequences of raising efficiencies. In keeping with the literature on the Jevons Paradox, I have retained the term 'paradox' for all four because the consequences for consumption (the left-hand side of Figure 37) and production (right-hand side of Figure 37) cannot be imputed solely from the efficiency ratio (Ruzzenenti and Basosi, 2008). In other words, these are 'paradoxical' because greater efficiency results in greater appropriation (rebound) counter to the notion that efficiency drives 'real' (aggregate) resource savings.

The first type of paradox arises because final products are more numerous and cheaper, reducing price and raising the consumer attraction and purchasing power. Polimeni *et al.* (2008) introduce this as efficiency-induced consumption of outputs, and Sorrell and Dimitropoulos (2008) term this 'direct rebound' (for specific produced goods) and 'indirect rebound' (for other general goods).

Second, the output may a have facilitative or power function (e.g. electrification, communication, transportation, miniaturisation) for accessing more resources. This is discussed as a powering of an 'economy-wide rebound' (Ruzzenenti and Basosi, 2008; Sorrell and Dimitropoulos, 2008) driven in part by a 'time saving' per unit of production.

Third if and when different pathways for losses exist (covered in Section 4.7.2), recovered losses may not recycle back to the natural capital pool but to its consuming 'owner' (proprietor) or to other users leading to increased total consumption of resources. This example underpins the recoverable yet subsequently consumed fraction of water in irrigation systems, and the great concern regarding the consumptive consequences of raising irrigation efficiency (see Ward and Pulido-Velázquez, 2008).

Fourth, claims that offsetting wastes and wastages to another location reduce aggregate consumption must be thoroughly tested bearing in mind the problem that offsetting might geographically transgress boundaries and scale but quantitatively fail to lead to a total reduction (Corbera *et al.*, 2010; Galik and Johnson, 2009). Furthermore, to switch production to another more efficient site as an argument to offset production at an original inefficient site without closing down the former is to invite additional control problems and a propensity to raise total production and consumption.

The productive-consumptive liminal space applies a variety of agro-industrial conversion processes probably best exemplified by the conversion of hydrocarbons such as coal or oil to useful energy output in kilowatt hours, as takes place in electricity generating stations and delivered to substations. A longer 'chain' of the conversion process might see the final product as electricity consumed in the household, allowing for demand side efficiencies to be analysed. Because of a lack of space, this type of paracommons is not discussed in detail.

5.2.3   Multipath liminal paracommons – irrigation in river basins

Building on Section 4.7.2 where the idea of different resolutions for the utilisation and saving of resources, wastes and wastages were discussed, the multipath liminal paracommons arises because of the high numbers of pathways combined with difficulties in controlling the exact routing of these pathways (or other similar resource). The multipath paracommons can be thought of as a special case of the productive-consumptive space except marked by an unusually high number of options that conversions and losses can 'decompose' to. In the case of irrigation, 16 pathways are identified – see Tables 20 and 21 and Figures 38 and 39. Figure 39 shows a geographic plan view of the pathways in Figure 38.[1]

By breaking the inflows, pathways, dispositions and destinations into stages, it is now possible to explore how the multipath liminal space arises. Figures 38 and 39 show freshwater coming into a proprietor system 'decomposing ' into different destinations by switching between a very large number of combinations available including options for forestalling, recovering, avoiding, offsetting and transferring the extrinsic fraction. Thus Figure 38 shows both the extrinsic and intrinsic fraction entering the irrigation system. From the intrinsic fraction flows the beneficial process fraction (being synonymous). From the extrinsic fraction flows any one of many pathways resulting in both intermediate stages

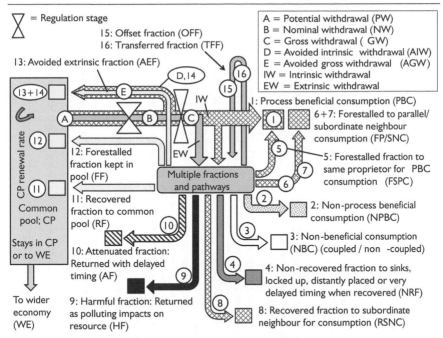

Figure 38 Multiple pathways and fractions of water in irrigation systems

(e.g. paragains to the proprietor or neighbour) and ultimately in the depletion or conservation of the common pool resource. Figure 38 also shows that some of the intrinsic fraction can switch to becoming an extrinsic 'loss' fraction. Furthermore, it is worth noting that the offsetting and transferring options (where losses are reworked elsewhere) cannot be understood to have 'as yet' decomposed into these outcomes and thus are pending; Figure 38 shows them remaining undecided as to what their final outcomes are. In Figure 38, one can think of all of these options as emanating from a transitive space of options – hence the term 'liminal multipath paracommons'.

The previous paragraph can be broken into more detail with the aid of Table 21. The far left- and far right-hand columns (A and F) of Table 21 give the inflows and final outcomes of fractions of water of a multipath paracommons exemplified by irrigation. Column A dissects the inflow into intrinsic and extrinsic fractions – the latter comprising the 'losses'. Column F, presents the final three dispositions: depletion, common pool retention and fate pending.

Between Column A and Column F sit a variety of stages. Column B describes efficiency/inefficiency problems visible on an irrigation scheme – such as standing water on uneven fields, or seepage at the end of the fields into drains. Column C gives a list of tangible actions that can be undertaken that generally but do not specifically bear down on losses and inefficiencies individually and exactly. Examples are levelling fields or lining irrigation canals. At the bottom

Table 20 Pathways and fractions of resources when irrigation systems are reworked

Main type	Sub-types	
Natural capital/ Common pool withdrawal	Potential withdrawal (PW) [A]. The amount of resource that can be harvested or withdrawn. It can exceed a water right or agreed amount. When exceeding the harvest capacity of a renewable resource, 'mining' is taking place.	
	Nominal withdrawal (NW) [B]. This is the fraction intended to be appropriated from the common pool that includes intrinsic and extrinsic fractions. It is defined by an established right or officially sanctioned technological constraint.	
	Gross withdrawal (GW) [C]. This is the water that is withdrawn at any given time, averaged over time. In some instances this will be greater than the nominal withdrawal but below or equal to the potential withdrawal. At other times, GW will be less than NW via pricing, a drought or water right changes.	
	Avoided intrinsic withdrawal (AIW) [D]. This is the intrinsic beneficial consumption that is not withdrawn into the system because it has been forestalled by a combination of substitution, dematerialisation and retrenchment.	
	Avoided gross withdrawal (AGW) [E]. When GW<NW, AGW equates to water subsequently not abstracted. As it includes intrinsic and extrinsic fractions, avoided withdrawal reduces losses ('Avoided coupled fraction' below).	
	Intrinsic withdrawal (IW) [1]. Synonymous with process beneficial consumption (PBC).	Extrinsic withdrawal (EW). This includes the below fractions numbered (2 to 16).
Consumed fractions (lost from the hydrological system)	Process beneficial consumption (PBC) [1]. This is the intrinsic resource consumed and/or converted to or locked up in the intended good. An example is crop transpiration.	
	Non-process beneficial consumption (NPBC) [2]. This is used beneficially but not in the production of the good. An example might be trees lining a canal.	
	Non-beneficial consumption (NBC) [3]. This is consumption from the hydrological cycle that has no beneficial or harmful impact (e.g. evaporation from open water in cropped fields).	
	Non-recovered/able fraction (NRF) [4]. These are losses that pass to sinks such as deep aquifers or the sea and represents water lost to further use.	

Losses forestalled or recovered then beneficially consumed	Forestalled to same proprietor consumption (FSPC) [5]. This fraction is reduced, released and used by the proprietor for beneficial consumption elsewhere.
	Forestalled to parallel neighbour consumption (FPNC) [6]. This fraction is economised by one user and made available to another user within the same level for beneficial consumption without delay.
	Forestalled to subordinate neighbour consumption (FSNC) [7]. This fraction is economised by the proprietor for a user next in the sequence for beneficial consumption without delay.
	Recovered to subordinate neighbour consumption (RSNC) [8]. This fraction is recovered by another user for beneficial consumption without delay.
Non-consumed fractions (not lost to hydrological system)	Harmful fraction (HF) [9]. These are wastes that pollute and degrade users or the common pool. They may offer some sustenance for specialised ecologies.
	Attenuated fraction (AF) [10]. This water returns to the common pool but is mildly polluted or delayed in timing as happens in irrigation systems.
	Recovered fraction to encasing system; (RF) [11]. Water returned to common pool or wider economy with little timing delay or quality loss.
	Forestalled fraction (FF) [12]. These are losses 'reducible' within the system that can be identified, acted upon and retained in the common pool.
	Avoided coupled fraction (ACF) [13]. These are losses coupled with intrinsic withdrawal and are released by avoiding withdrawal (via substitution, dematerialisation and retrenchment).
Pending fractions (fate uncertain)	Offset fraction (OFF) [14]. Options for offsetting losses require improvements to efficiency elsewhere in the river basin.
	Transferred fraction (TFF) [16]. Options for switching production to another location possibly incorporating reductions in losses.

Table 21 Transitional/liminal framework of multipath paracommons

Inputs and fractions in withdrawal A	Vernacular 'visibility' of irrigation losses (selected examples) B	Multipath liminal space			
		Water control technologies examples C	Fractions as pathways; options for switching in column E D	Fractions as portions E	Fractions as dispositions F
Intrinsic withdrawal (IW) [1]	• Crop transpiration	• Canal system management • Canal system leakage • Canal weeding • Canal density and number		Process beneficial consumption fraction (PBC): water evaporated or transpired for the intended purpose by the proprietor [1]	Depletion. Lost to effective use or locked up or polluting.
All other extrinsic withdrawal fractions (EW)	• Conveyance spillage and leakage • Bare soil evaporation at field edges • Canal evaporation • Canals with silt and weeds • End of field drainage • Weeds transpiration • Excessive standing water • Seepage below roots • During session canal leaks to field and drains	• Canal flow control technology • Hydromodule/water duty • Field and in-field design, e.g. gradient, basin or furrow morphology • In-field walkways • Deficit scheduling • Crop season length • Field ploughing • Field pre-watering • Crop selection and patterns	Main pathways and switches: EW → PBC (EW→1) EW → NPBC (EW→2) EW → NBC (EW→3) EW → NRF (EW→4) EW → FSPC (EW→5) EW → FPNC (EW→6) EW → FSNC (EW→7) EW → RSNC (EW→8) EW → HF (EW→9) EW → AF (EW→10) EW → RF (EW→11) EW → FF (EW→12) EW → AEF (EW→13) EW → OFF (EW→15) EW → TFF (EW→16)	Non-process beneficial consumption [2] Non-beneficial consumption (NBC): water evaporated or transpired for purposes other than intended use [3]. Also possibly offset [13] and consumed elsewhere Non-recovered fraction (NRF): Water that is lost to further use e.g. sinks [4]	

	Other examples of switches:		
• End of session draining of canals to drains • Application of irrigation after crop has wilted (evaporation not transpiration)		Released consumed: Water consumed by proprietor, parallel or subordinate [5, 6, 7]	
• Command area control • Accommodating rainfall	RSNC reduced → BC (8→1) RSNC reduced → NBC (8→3)	Recovered consumed: Water is recycled and consumed by subordinate user [8]	
Requiring: • Farmer groups • By-laws • Agreed targets • Monitoring • Budgets • Sanctions • Meetings and training	RSNC reduced → FF (8→12) NBC reduced → PBC (3→1) NBC reduced → RF (3→11) HF reduced → AF (8→10) PBC reduced → NBC (1→3)	Harmful fraction (HF): Polluted water degrades common pool [9]	
		Attenuated fraction (AF) [10] is delayed in returning to common pool	
		Recovered fraction returned to CP [11]	Retention in CP and/or WE
		Forestalled fraction (FF) [12]	
		Avoided extrinsic fraction (AEF) [13] Avoided intrinsic withdrawal (AIW) [14]	
		Offset fraction (OFF) [15]	Fate pending
		Transferred fraction (TFF) [16]	

(Note: abbreviations, letters and numbers correspond to Figure 38.)

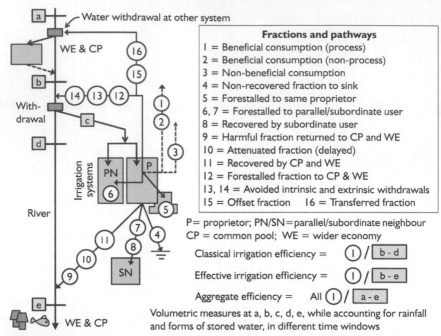

Figure 39 Irrigation fractions as pathways in geographic space

of Column C are other institutional interventions that influence technologies chosen. Columns D and E are part of the complicated and 'messy' liminal space where water takes or switches between various pathways. Thus the technical options in Column C give rise to switching options in D, resulting in changes to fractions in Column E leading to outcomes in column F.

The reason Columns C to E are 'liminal' is because in trying to raise efficiency from a current situation to a future more efficient condition, it is extremely difficult to discern and predict how interventions listed in Column C (to deal with the visible problems in Column B) precisely secure outcomes in columns D, E and F, especially when the latter is also expressed at the river basin scale. Furthermore, the conditions in Column B have to be judged in terms of how deleterious they are to overall efficiency and what future condition is sought. For example, farmers might discuss how bad a particular canal leak is and what action is required: to be left, patched or fully repaired. The indeterminate problem here is that the newly salvaged once-leaking water does not then assuredly move to beneficial consumption. This water could be lost elsewhere – even at the original canal leak if poorly repaired. Irrigation systems covering many hundreds of hectares present many opportunities for various fractions and outcomes to imprecisely merge into each other. If the effort to raise the performance of a complex system can be characterised as never-ending Sisyphean task, the irrigation system and its occupants are in a permanent state of being situated 'in between' a set of views about today's efficiency (expressed in different ways as Columns A to F shows), expectations regarding efficiency in the future, and selecting interventions that then play out in unexpected ways. (I consider there to be similarities here to work conducted by anthropologists (Lewis, 2009; Li 2007) on the promise of a better future delivered by international and community development in the face of overlapping and externally applied objectives combined with inexact knowledge of the system being addressed.)

In summary, attempts to improve irrigation efficiency are undermined by control and scale discontinuities between the columns and fractions, as well as opinions regarding current conditions and future options for each of the columns, as well as other issues identified in the REC framework such as ill-defined terms and definitions, coupling between fractions and incomplete measurement.

Thus while the 'commons' describes the main canal water, crop transpiration and drainage water as real flows in real localities (in other words as 'places'), the paracommons describes the relationship between them, with the irrigation system and its management forming the transitive threshold between options to adjust where resources and losses move to. Behind these options lie many more switching options (column D), and behind these sit a wide choice of technical, tangible activities (column C). Although Perry (2007) identifies recovered flows he does not provide options to take water back to the hydrological cycle and the wider economy, or water to different types of neighbouring users or for avoiding or forestalling 'losses'. It is this collection of multiple possibilities that generates the 'in-betweenness' or liminality of the paracommons.

Example calculation of paragains

The framework of pathways and fractions in Table 20 can be utilised to compute paragains for a proprietor irrigation system moving from a current 'inefficient' status to a future 'more efficient' condition (Table 22). The two messages from Table 22 is that a more detailed resolution of various pathways allows for a more accurate assessment of the movement of salvaged resources from sources to destinations and that different ways of conceiving of salvaging losses allow for different ways of calculating paragains (Table 22 is different from Table 13 which in turn is different from Table 4). To repeat, this worked example is attempting to demonstrate how a salvaged efficiency gain can be distributed to one or more of the four parasystem destinations depending on how the inefficient component of the resource is altered over time using the 16 pathways in Table 20.

The geographical spatial map of Figure 39 can be used to understand Table 22 – except that two pathways are not employed: offsetting efficiency (15) and transferring production (16) to another system elsewhere in the river basin. Instead, in this example the parallel and subordinate neighbours are directly connected to the proprietor system (PN, SN and P respectively in Figure 39). This makes Table 20 a different kind of example from that covered in Table 13

Table 22 Alternative calculations of paragains

		Time frames		Paragain		Net loss or gain
		Today	Future	Source	Disposition	
Nominal withdrawal (NW)		400	400			
Assumed efficiency to calculate avoided fractions		60%	60%			
Avoided intrinsic withdrawal (AIW)		30	60			
Avoided coupled fraction (ACF)		20	40			
Avoided gross withdrawal (AGW)		50	100			
Gross withdrawal into proprietor system (GW)		350	300			
Depleted proprietor losses or extrinsic depletion	Non-process beneficial cons [2]	10	5	−5		
	Non-recovered fraction [4]	50	20	−30		
	Non-beneficial consumption [3]	55	20	−35		
	Harmful fraction [9]	15	10	−5		
	Attenuated fraction [10]	45	20	−25		
	Sub-total	175	75			
Proprietor beneficial consumption	Net BC [1]	150	140	−10		
	Forestalled/retained BC [5]	0	25		25	
	Sub-total of BC	150	165			15
Neighbours' beneficial consumption (from proprietor)	Forestalled to parallel BC [6]	0	10		10	
	Forestalled to subordinate BC [7]	0	10		10	
	Recovered to subordinate BC [8]	20	10		-10	
	Sub-total	20	30			10

Common pool and wider economy					
Avoided intrs. withdrawal (AIW) [13]	30	60		30	
Avoided coupled fraction [14]	20	40		20	
Forestalled fraction [12]	0	15		15	
Recovered to CP and WE [11]	5	15		10	
Sub-total	55	130	−110	110	
Total	400	400			75
Aggregate BC consumption [1, 5, 6, 7, 8]	170	195			
Total proprietor consumption [1, 2, 3, 5]	215	190			
Depleted proprietor losses (from above)	175	75			
Aggregate consumption (1, 5, 6, 7, 8, 2, 3)	235	220			
Aggregate depletion (Agg BC + depleted prop loss)	345	270			
Salvageable losses resolved [5, 6, 7, 8, 11, 12, 14]	45	125			
Total extrinsic fractions resolved	220	200			
Final proprietor CIE (= prop BC/GW)	42.9%	55.0%			
Final proprietor EIE (= prop BC/total prop consumpt)	69.8%	86.8%			
Parasystem aggregate CIE (= Agg BC/GW)	48.6%	65.0%			
Parasystem aggregate EIE (= Agg BC/agg consumpt.)	72.3%	88.6%			
Total system paragain (today agg depl – future agg depletion)	75	75			

where two systems X and Y, each with their own point of water withdrawal from the river basin, were portrayed. Table 22 shows changing fractions for two time frames 'Today' and 'Future'. Then in the column to the right of these (labelled 'Source'), the difference between 'the more efficient future' and the 'less efficient today' shows how the size and location of the paragain is computed from changing fractions. Finally, the calculation of the paragain resolving to different 'dispositions' (e.g. beneficially consumed or returned to the common pool) and the computation of final net gain (or loss) for the destination party are given on the right-hand side of Table 22.

In Table 22, the nominal withdrawal by the irrigation system in both situations (today and future) is 400 units. However this nominal amount is not withdrawn because of avoided gross withdrawal (made up from an avoided intrinsic withdrawal and an avoided coupled fraction). The future system increases its avoided gross withdrawal (AGW) to 100 units (from 50 in the current situation) giving a future gross withdrawal (GW) into the system of 300 units (compared to 350 units of GW currently). The expected classical efficiency of 60 per cent allows this AGW to be split into an intrinsic and extrinsic amount of 60 and 40 units respectively for the future system. (This calculation of avoided withdrawal allows 'savings' to be made by not withdrawing an intrinsic water fraction that has losses coupled with it.)

The next five rows subtract from the newly determined gross withdrawal (of 350 and 300 units) the depleted extrinsic fractions that are deemed to be associated with the proprietor system. In the future more efficient system, these amount to 75 units rather than 175 units in the current system. These extrinsic fractions are subsequently resolvable into different beneficial (intrinsic) fractions by being forestalled. In the future scenario the forestalling and reduction of these 'losses' allow larger BC fractions for the proprietor and neighbour as explained next.

Then there are two rows which allow the computation of the beneficial consumption by the proprietor. The first row is the amount of water intended to be consumed by the proprietor, which in this case drops from 150 units 'today' to 140 in the future (The second row is that which is forestalled and released by a reduction in losses. This amount is greater in the future (25 units up from zero units) giving a total proprietor beneficial consumption of 165 units. The 150 units of total BC today can be calculated in the table as the GW (350) minus extrinsic proprietor losses (175) minus recovered to neighbouring users (20) minus recovered to the common pool/wider economy (5). The same calculation is not applied to the future column because fractions resolve to different destinations. In the future case the 165 units of total BC can be calculated in the table as the future GW (300) minus extrinsic proprietor losses (75) minus recovered to neighbouring users (30) minus recovered to the common pool/ wider economy (15) minus forestalled to the CP/WE (15).

The next set of rows describes how neighbouring systems (parallel and subordinate) pick up water from the proprietor system by being released and by being forestalled by the proprietor. In the more efficient future, neighbourly beneficial consumption increases to 30 units from 20 units 'today' (although this includes a drop in the amount of water recovered from 20 to 10 units).

Next, the calculations for the combined common pool (CP) and wider economy (WE) are derived. There are four ways that water is retained in the CP/WE: avoided intrinsic withdrawal, avoided coupled fraction (which together add up to the avoided gross withdrawal), forestalled losses and recovered losses. In the future the CP/WE, via these four pathways, retain and receive 130 units as compared to 55 units under the current scenario.

Using Table 22, it is now possible to draw attention to various dimensions of the irrigation efficiency debate (bearing in mind that Table 22 only includes 14 of the 16 pathways):

- A focus only on recovered fractions (RF) for a basin-level model (Perry, 2007, 2011) would give 5 units under the current scenario and 15 units under a future scenario, and thus omit the many other ways water can be salvaged and redirected to new destinations.
- Furthermore, a focus only on aggregate depletion between the two scenarios offers a limited picture of how much water is salvaged. Using only aggregate depletion, the paragain size works out to be 75 units (current aggregate depletion = 345, future = 270). (Where, aggregate depletion is the sum of all consumption and all extrinsic fractions not retained in or recovered by the common pool.)
- Thus a focus only on salvaging losses (e.g. as in Table 13) without specifying their destinations means that the 'ins and outs' of resources and losses flowing between the proprietor and three other destinations are not revealed. For example in Table 22 the proprietor gains a net total of 15 units (she reduces BC from 150 to 140, but forestalls and retains an extra 25 units of 'losses'). The neighbour gains an additional 10 units from the Proprietor, and so increases beneficial consumption from 20 to 30 units. And the CP/WE gains 75 units (calculated via the aggregate depletion changes in the previous bullet point) largely by avoiding withdrawal but also by a forestalled fraction of 15 units retained in the CP and a greater amount of recovered losses, up from 5 to 15 units).
- Much more complicated models, going beyond units, could be created incorporating hectare areas, mm depth equivalents, volumes and flows, as well as explicit options for offsetting efficiency and transferring production, plus allowing variables for substitution, dematerialisation and retrenchment.

In summary, the size and nature of the paragain is wrapped up the accounting model selected including its degree of resolution regarding how to make savings, the start and end points, the source they come from, the type of neighbouring systems, the salvaging means employed and the destinations they travel to. This is why the gain is arguably subjective and why the prefix 'para' is employed. The sense of future alternatives juxtaposed against current assumptions is important because the recovered or forestalled loss is either a resource for its 'owner', or for other claimants such as neighbouring irrigators or state interests such as hydropower and maintaining ecological flows. Therefore the intention to switch resources between pathways can, via recycling or losses foregone, become associated with claim and counter-claim over ownership by users who are accustomed to current pathways and distributions of resources and losses.

5.2.4 Multiservice liminal paracommons – carbon, forests and soils

This section applies a paracommons view to the trade-offs between and distributions of ecosystem services (ES) following interventions designed to manage ES (Gordon et al., 2010; Bailey *et al.*, 2011; Leach *et al.*, 2012; Sikor, 2013; Muradian and Rival, 2013). In some senses nothing new is being proposed here because ES theory and scholarship already tests ideas of equity and justice when one ecosystem service is boosted to the detriment of others (in other words 'who gets the gains of an ecosystem emphasis').[2] Nevertheless, despite there being limited space to explore the jostling of, and changes to, ecosystem services alongside each other, the paracommons appears to equate more to an ecosystem services view of natural resources than a 'commons' framing might, as is now explained.

In the multiservice paracommons, conversions of common pool elements and compounds are seen as ecosystem services (following on from Section 4.7.3). 'Multiservice' describes these qualitatively different pathways. An example is carbon which is found as an atmospheric gas carbon dioxide, subject to conversions into vegetative growth and subsequent services pathways into forest products, crop production, crop residues and soil organic matter, biochar and solutes. The multiservice paracommons explores the impacts of the efficiency of increasing certain services over others; in other words, whether, say, carbon sequestration can be boosted over timber production through efficiency amd productivity actions and whether this means, that by capturing more carbon into one service, alternative carbon services are reduced or cut off.

Thus, carbon-based ecosystems services comprise a complex and transitive field because of the number of services and flows functioning alongside each other. For example, Law and Harmon (2011) identify 22 flows between atmosphere, biofuel stock, long-term products, forest C stocks, short-term products and landfill. Lal (2008) counts 16 different sequestration pathways to lock away carbon. Importantly, these forms in turn influence other desirable or non-desirable soil and catchment behaviours and ecosystem services such as biodiversity, hydrology, runoff regimes and aquifer recharge. Adding further complexity, the performance of long-term sequestered carbon in forests is offset against the production of fossil fuel carbon. Law and Harmon (2011)

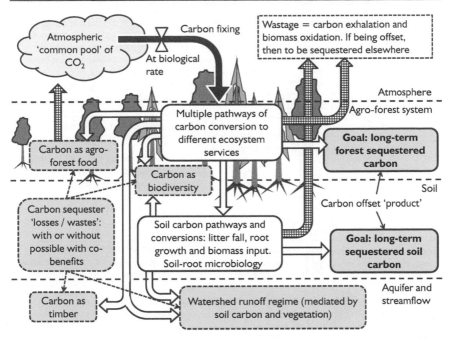

Figure 40 Multipath paracommons: ecosystem services of carbon conversions

explain these benefits in their paper, as do other articles that surround REDD+ (e.g. Thompson *et al.*, 2011). These flows, switches, adjustments, intermediate forms and consequent services indicate the presence of options and stages which in turn suggests that a liminality or threshold exists between primary or basic elements and final 'service' outcomes.

The multiservice paracommons is now more closely examined via an example of carbon sequestration. The efficiency/productivity problem in Figure 40 sees a world where, in order to reduce atmospheric carbon dioxide, carbon sequestration is the prime objective. The problem is: 'how might the carbon dioxide 'common pool' be purposively managed via photosynthesis to more productively achieve long-term sequestered carbon operating in the shadow (alongside) of other ecosystem services such as forest biodiversity, and hydrological runoff?' The complexity problem for this question is how to guide conversions towards prioritised pathways rather than to 'losses' or other less desirable ecosystem services (less desirable if long-term carbon sequestration is the main goal) or wastage of carbon dioxide back to the atmosphere. Recall, the term 'losses' is employed when carbon turns up in services other than in the deemed priority ecosystem service (terms such as costs or opportunity costs describe ES foregone in the REDD+ literature; Corbera *et al.*, 2010; Hufty and Haakenstad, 2011). This question thus asks how might a 'supporting ES' (carbon sequestration) be boosted in the face of losses in trying to achieve that prioritised ES.

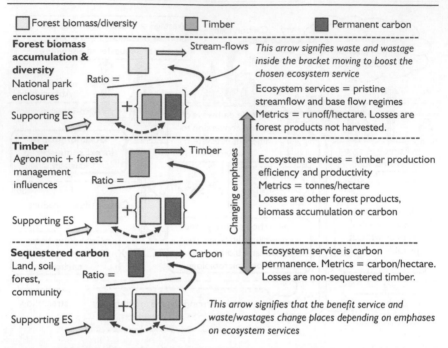

Figure 41 Changing the numerator emphases of ecosystem services

In a first analytical step, it is relevant to discern what compromises goods, wastes and wastages. Figure 40 reveals that the prioritised 'service' is long-term sequestered carbon. Thus, carbon dioxide is both the common pool 'source' and the wastage product returning to the common pool (by oxidation of short-term carbon, for example, rotting leaves and food). This means short-term fixed carbon becomes the 'waste' product in seeking this goal, but which nevertheless produces other important beneficial services (such as leaves, food, biodiversity and streamflows).

To further probe the efficiency dimensions of Figure 40 and the multiservice paracommons, Figure 41 show how priorities create desired services and less desired services. Figure 41 offers three main carbon-based services: forest biomass and biodiversity, timber production and carbon sequestration, recalling that is the latter that the forest is to be manoeuvred towards in our worked example.

At the top of Figure 41, the main emphasis sees forests as pristine environments to conserve biodiversity and boost biomass accumulation which in turn influences streamflow regimes. In the middle, repeated timber harvesting becomes the emphasised ecosystem service, which reduces the significance of biodiversity and accumulated biomass. At the bottom, permanently sequestered carbon becomes the emphasis which demands different types of forest and timber management. Figure 41 therefore captures two types of management decisions;

a) which ecosystem service gets the emphasis, and then b) how the productivity of that emphasised ecosystem service is boosted and increased. Figure 41 also implies a third matter falls out of these two decisions: what happens to the ecosystem services now 'downplayed' (in other words the deemed 'losses')? To explain this, five types of arrows are used in Figure 41:

- First there are three white arrows on the left-hand side coming into the denominator of the efficiency ratio which can be thought of as all potential supporting ecosystem services and natural inputs. These are nutrients and nutrient cycling, soil formation processes (e.g. pH), sunlight, water, gases and so on. These 'incoming' or base supporting ecosystem services give rise to higher services manifesting themselves as provisioning, regulating and cultural services that we, as humans, are interested in. However, at this point in the denominator of the efficiency ratio, they sit as potential ecosystem services for a given tract of land.
- On the output side there are three differently shaded arrows, one for forest biomass and biodiversity, one for timber and one for carbon sequestration. These are the services selected and then boosted that emanate from the prioritised numerator.
- Then there is a large vertical arrow that straddles all three services. This denotes selection over which ecosystem service is selected/emphasised prior to the second objective of boosting the service (discussed below).
- After the decision to emphasise one ecosystem service over another, the black solid arrows reflect attempts to manage the service. These 'take' the supporting services that are situated in the denominator and move them to the numerator. In other words the black arrows represent a set of managerial decisions that boost the off-take or productivity of the potential ecosystem services available such as biomass accumulation, timber offtake or carbon sequestration.
- Finally the black dotted arrows at the bottom of the three denominators explain how the 'lesser important service' (in other words the 'loss') are identified within the denominator. For example, moving from biodiversity to timber, or boosting the latter, switches not only the service (the solid black arrow) but the nature of the 'losses' (the dotted arrow).

Summarising Figure 41, there are two different overt managerial steps: selecting/emphasising the ecosystem service and then increasing the 'rate' of the service. This then leaves the third implicit, residual or 'hidden' matter of losses. These are taken in turn, although for the sake of brevity and clarity, a more nuanced detailed discussion has not been possible.

First, the switching between ecosystem services primarily arises from choices regarding land-use planning. While the contemporary story of turning a pristine rainforest into an oil-palm plantation offers a stark example of this switch (Fitzherbert et al., 2008), debates over land-use priorities have a long history. A fascinating article from the 1970s identifies the tensions facing the UK Forestry Commission regarding its priorities: 'objectives for its forestry enterprise in certain forests of the national parks, whereby the production of timber would become secondary to the protection and enhancement of the environment and the provision of recreation facilities' (Towler, 1975, p.133). He goes on to explore how this change in emphasis might play out with neighbouring private woodland owners.

Second, to boost the efficiency and productivity of an ecosystem service, it remains important to discern the extent to which 'natural' conversion processes (and the ecosystem services they influence) may be adjusted, depending on our wish for which ecosystem services is to be 'turned up' e.g. for more carbon to be sequestered (Mueller *et al.*, 2004; Nelson *et al.*, 2008; Antle *et al.*, 2003). While the productivity rate of carbon fixing may be argued to be a largely unalterable natural process or that nature is always 'efficient', it is possible to briefly outline the ways in which long-term and verifiable carbon sequestration (to select one service) may be steered and increased (Ott *et al.*, 2011). Some of the options (see Ryan *et al.*, 2010) include: land use choices regarding primary or secondary forest or agroforestry or other crops; crop choice involving C4 and C6 crops; timber species choice; harvesting frequency for timber products; biochar management, and soil drainage for soil organic matter oxidation.

Third, it is now relevant to ask 'What is the extrinsic loss fraction that can be salvaged within ecosystem service?' In other words, what is a carbon-service paragain and who gets this when trying to raise the efficiency of ecosystem services? This is difficult to answer because services are not subtractive as physical resources are. In other words, a reduction in soil carbon does not give an equivalent reduction on rainfall infiltration which in turn leads to an equivalent reduction in hydrological base flows. However, questions of paragains, destinations and neighbourliness are pertinent for the multiservice paracommons.

Thus Campbell (2013) describes how Nepalese villagers working and living in close relation to national forest parks previously accessing forest products, especially timber, are excluded from harvesting even low-value firewood due to changes in bureaucratic priorities over forest reserves in turn borne from national interpretations of international concerns regarding carbon emissions. The case shows that an overt priority towards reducing carbon emissions from forests means that the forest gets the gains from attempts to raise this kind of performance while the villagers are excluded from 'picking up' (figuratively and physically) the subordinated 'losses'. Aside from loss of livelihood and arrival of new forms of environmental subjectivity, the villagers' switch to manure as a source of energy as a consequence of this policy makes little sense quantitatively in terms of reduced emissions in the search for more eco-efficient (from a bureaucratic point of view) use of state parks. Naïve views by policy-makers of the partitioning of ecosystem services, while depressing, are perhaps predictable. Further discussion of paragains is found in Section 5.3.2.

Other aspects of the paracommons similar to the previous section can be discerned. The conversion process of carbon dioxide to long-term sequestered carbon is 'stretched' through multiple stages and made problematic via wider boundaries and longer time spans, each with opportunity for 'leakage' (Lal, 2008) where sequestration is not fully pursued to an aggregate verified locked away carbon. Law and Harmon (2011) also draw attention to time and scale in carbon accounting and Rice (date unknown) points out the many intervening variables that interfere with guiding prioritised and subsidiary carbon services to chosen outcomes.

Thus, as with the multipath space, the technologies and control required to switch and direct carbon through the multiservice space to precisely one of these final outcomes is likely to be problematic and vague (Galik and Jackson, 2009). In addition these choices must be framed, invariably problematically, by opinions about poor sequestration performance today and better sequestration performance in the future. As Ryan *et al.* (2010) point out on page 4 with reference to a graph showing the dynamics and average of carbon stored over time:

> Management actions should be examined for large areas and over long time periods. This figure illustrates how the behaviour of carbon stores changes as the area becomes larger and more stands are included in the analysis. As the number of stands increases, the gains in one stand tend to be offset by losses in another and hence the flatter the carbon stores curve becomes.

This makes the point that scale and time horizons define the boundaries (recognising, for example, the risks to carbon build-up by fire) by which bona fide sequestered carbon can be assessed. Concerns voiced by Leach *et al.* (2012) about biochar permanence also point to questions of competing land uses and an industrial scale of 'production' to achieve significant impact.[3] The authors point out questions of future distribution and justice regarding 'whose lands are selected' for biochar production. This putative switch in land use means that other ecosystem services upon which people's livelihoods rest (firewood and livestock) are demoted as wasteful – all framed by the elevated and untested (at the landscape scale) promise of biochar to significantly reduce net CO_2 levels.[4]

In Figure 42 I contrast a conventional commons and liminal paracommons framing of forests. In the former (the top of Figure 42), forest communities are concerned with the renewable and renewing biology of 'their' forest reserve, mostly likely on a seasonal to two or three year time horizon. The productivity problem is one of matching harvest rates with biological production of forest and forest products. Here, the common pool rights challenge is to exclude free-riders and over-harvesting. In the bottom half of Figure 42, forest parasystem communities now have to consider many more trade-offs in addition to their commons approach to 'their' forest including local and non-local services: long term carbon sequestration (beyond 50–100 years); offsets for other carbon dioxide

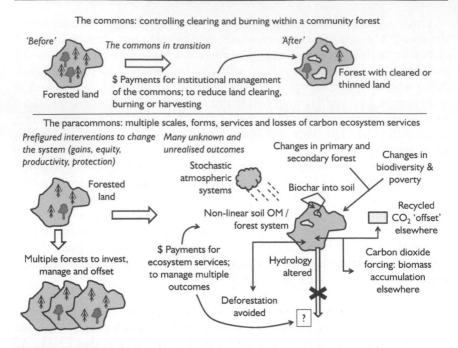

Figure 42 Contrasting the commons and paracommons (forestry and carbon)

sources and conversions; consequences for the global atmospheric commons and impacts on other ecosystem services such as biodiversity and downstream river flows. Here the rivalrous challenge is how to recognise and give space to deprioritised services in a forest system overtly seeking newly prioritised services. The betwixt-and-between liminality operating here tells us that endeavours to create these more equitable, beneficial and productive versions of the erstwhile forest commons will be immeasurably subject to greater degrees of freedom and nth variables suggesting outcomes may differ very greatly from expectations.

In the case of Nepal highlighted by Campbell (2013), mentioned earlier, seeing all forest biomass as an opportunity for preserving carbon stocks reveals aspects of the paracommons. It is calculatively and scientifically naïve (with a vanishingly small fraction of carbon likely to be permanently sequestered if not more robustly targeted); it extends government and governmentality; it is uninterested in monitoring carbon 'goods' (and subsequent losses), and it excludes villagers from being part of a debate on how to achieve multiple priorities and services side by side. This situation may be interpreted as a problematic habitation of the betwixt-and-between 'expectations' afforded to newly industrialising ecosystems with questionable accounting and monitoring to guide the system towards firmer resolutions. While the Nepal case is perhaps a stark one, my view is that the distributional/justice concerns of carbon-based ecosystem services partly arise out of a prior underestimation and misapprehension of the efficiency and productivity dimensions of those ecosystem services. In other words, initiatives will probably fail to deliver both substantial and verifiable numerator ecosystem services and compensate by salvaging demoted denominator ecosystem services. I read the concerns voiced by Leach *et al.* (2012) on biochar promise in that same light. Further discussion on ecosystem services is found in Sections 5.3.2 and 7.2.

5.3 The efficiency gain destinations of the three liminal paracommons

By utilising the four-vector quadrangle introduced in Figure 23 it is possible to interpret how different paracommons guide or steer the material gain of the efficiency gain. Three interpretations are selected to demonstrate the quadrangle; a) irrigation efficiency concepts via the multipath paracommons; b) ecosystem services via the multiservice paracommons, and c) selected productive-consumptive examples.

5.3.1 Multipath: three paradigms of irrigation system efficiency

The three approaches to understanding irrigation efficiency (fractions/EIE, classical or conventional approaches, and the paracommons) are discussed in terms of the destinations for salvaged losses (Figure 43).

Conventional and classical efficiency view of irrigation systems

I have placed a conventional view of irrigation systems towards the proprietor/ single system corner of Figure 43. This is for two reasons: first, classical efficiency, emphasising 'process efficiency', dominates analyses of individual irrigation systems, rather than effective efficiency which tests the impacts of irrigation on the river basin. Closely related, second, irrigation continues to be dominated by professional engineering viewpoints seeking to raise efficiency of particular systems by conventional solutions such as canal lining, metering and drip irrigation, rather than by participatory farmer-led solutions which might emphasise rotating water between fields, changes to cropping patterns or the use of different labour and land intensities (see Boelens and Vos, 2012). Although farmer-led solutions might enhance proprietor performance, a shift towards collective co-management might bring with it a re-balancing of emphasis between single and neighbouring systems which might take paragains along the neighbourly axis.

Effective efficiency and water accounting

In Figure 43, I have placed effective irrigation efficiency and water accounting towards the environmental/common pool corner with a stretch towards the

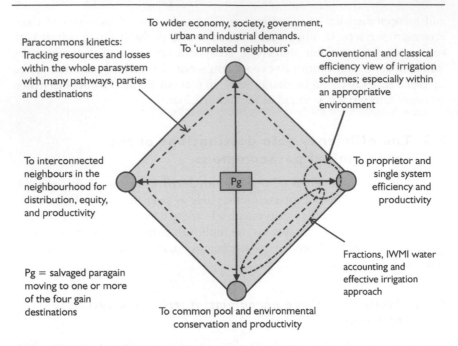

Paracommons kinetics:
Tracking resources and losses
within the whole parasystem
with many pathways, parties
and destinations

To wider economy, society, government,
urban and industrial demands.
To 'unrelated neighbours'

Conventional and classical
efficiency view of irrigation
schemes; especially within
an appropriative
environment

To interconnected
neighbours in the
neighbourhood for
distribution, equity,
and productivity

Pg

To proprietor and
single system
efficiency and
productivity

Pg = salvaged paragain
moving to one or more
of the four gain
destinations

Fractions, IWMI water
accounting and
effective irrigation
approach

To common pool and environmental
conservation and productivity

Figure 43 Irrigation efficiency models in the paragain quadrangle

proprietor performance corner. This zone of interest is a consequence of the reporting of final dispositions – essentially towards process and non-process depletion and outflows. For example, it is because the IWMI-WA does not report on the intricate detail of water and loss distributions to different users and their neighbours within one level or scale that this approach does not spread towards the neighbouring apex. It is also my view that Perry/IWMI-WA water accounting does not, in its current formulation, report on the various ways paragains are manoeuvred through the parasystem to different destinations.

Multipath paracommons

Demonstrated by irrigation in river basins, the key feature of the multipath paracommons is one of attempting to understand the distribution of paragains and the consequences this might have for equity and performance for all systems and parties involved. For this reason, this space occupies a broad region throughout the quadrangle in Figure 43 which denotes the parasystem. The multipath paracommons is interested in how the material gain is shared between the proprietor system and any given immediate neighbouring system, but also, via the river basin, to common pool conservation and the wider economy.

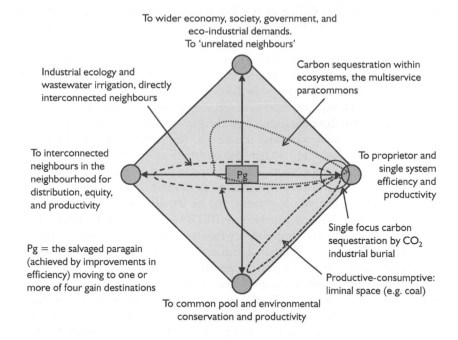

To wider economy, society, government, and
eco-industrial demands.
To 'unrelated neighbours'

Industrial ecology and
wastewater irrigation, directly
interconnected neighbours

Carbon sequestration within
ecosystems, the multiservice
paracommons

To interconnected
neighbours in the
neighbourhood for
distribution, equity,
and productivity

To proprietor and
single system
efficiency and
productivity

Pg

Pg = the salvaged paragain
(achieved by improvements in
efficiency) moving to one or
more of four gain destinations

Single focus carbon
sequestration by CO_2
industrial burial

Productive-consumptive:
liminal space (e.g. coal)

To common pool and environmental
conservation and productivity

Figure 44 Socio-ecological systems in the paragains quadrangle

5.3.2 *Multiservice paracommons in the efficiency quadrangle*

Another version of the parasystem quadrangle (Figure 44) explores how losses
are shared in the multiservice paracommons. The first approaches carbon
sequestration via ecosystem services, in contrast to an overt industrial process
discussed in the next section.

Carbon can be sequestered by reworking ecosystem services to boost, for
example, biochar production or timber utilised for the construction industry
(rather than as a fuel). Because carbon sequestration is the selected example
of the multiservice liminal space, this zone pulls towards the proprietor
performance vertex but also fills the quadrangle towards the neighbour vertex
and wider society vertex. The following points explain this pattern.

First and foremost, emphasis is placed on the performance and productivity
of carbon sequestration by the proprietor system (e.g. the privatised forest
company or project targeting carbon sequestration). Thus a more efficient
carbon sequestration will lead to those salvaged losses going to the owner of the
system – the proprietor – attempting to sequester carbon.

Second, however, in raising the efficiency of carbon sequestration there are
likely to be other ecosystem services generated that may have some element
of parasystem distribution reaching up to the top and left-hand vertices.
Thus within systems attempting to permanently sequester carbon, different

ecosystem services are affected by the different flows, locations and intensities of short-term carbon (the less important 'waste' in this system) in turn affected by the decisions taken over land, forests, labour, inputs, water and so on. For example, immediate neighbouring gains might be harvested by other forest users, and wider indirect neighbour services provided by improved river runoff if afforestation delivers this alongside the target of carbon sequestration.

Thus this parasystem distribution of services subsequently disproportionally impairs and benefits different users depending on the focus and effort applied to streams of carbon-based goods and services as they cascade through the forest/soil/hydrogeology system and are felt both within the forest by other users of the forest (neighbours) but also wider afield in the catchment (the wider economy). Furthermore, it is because that sequestration performance and process efficiency is the prime focus, that whole systems can be transferred or offset from one location to another. This is largely how REDD+ functions, deciding which projects to invest in. This compensatory mechanism of switching productivity and process efficiency to different locations subsequently determines which forests (and users) reap or lose from locational switching.

Note, however, that the ecosystem-based carbon sequestration is not drawn towards the bottom vertex of Figure 44, that of common pool conservation. This may seem counter-intuitive but recall, in attempting to raise the efficiency of sequestration, CO_2 preservation in the common pool is not sought. In other words, having turned the carbon commons upside down, the preservation of carbon dioxide in the 'common pool' atmosphere is not the aim of a carbon sequestration system. Although there is a broader environmental/climate change benefit to sequestering carbon, it is the nature of the common pool, goods and services that determines the location of the salvaged ecosystem services moving through the quadrangle.

5.3.3 Productive-consumptive paracommons

Three examples of the productive-consumptive paracommons (also in Figure 44) provide a contrast on the significance of the 'source' or common pool material. In the first example, the common pool is coal, in the second it is atmospheric carbon dioxide and in the third it is one factory generating waste for another or an upstream city issuing wastewater.

Carbon (coal) preservation by mining less coal

I interpret the original 'coal question' of the Jevons Paradox as a productive-consumptive space inhabiting the lower part of the quadrangle between environmental protection and productivity performance. This is because although the aim is to produce more energy from coal more efficiently, systems defined by this liminal space have a paradoxical tendency to result in more natural capital depleted. Getting this balance right is the aim of this space.

Carbon sequestration by burying carbon dioxide

Here the aim is to sequester carbon via a purely industrial-geological process with zero interest in preserving the common pool of carbon dioxide. If the efficiency of this process is raised, then the 'losses' of CO_2 back to the atmosphere are reduced with the consequence that the salvaged gain in sequestered CO_2 is put into rock strata – boosting the success and performance of the proprietor company owning this technology. Hence material gains in efficiency 'move' to the proprietor company – the right hand apex of the quadrangle.

Industrial ecology and wastewater irrigation

Conceived simply, industrial ecology is interested in how one factory provides the inputs to another factory, creating a symbiosis by this transference of wastes as resources. A similar relationship exists between an urban centre and its peripheral wastewater irrigation dependent on effluents from the city. Although I do not see them as full paracommons systems, I have placed both within a zone stretching along the neighbourly axis between the proprietor system (the first factory and city, respectively) and their neighbouring system (the receiving factory and peripheral irrigation system). This reflects the idea that any gain in efficiency in the proprietor system (not easy to define in this case) can be passed to its symbiotic partner. Conceptually, the donating sector could be viewed as the common pool, but here the heightened immediacy of the connection between the two systems suggests that industrial ecology falls along the proprietor-neighbour axis.

5.4 Chapter review

The chapter introduced liminality – or transition over a threshold – as a key defining feature of the paracommons. Liminality arises because efficiency entreats systems and their managers to move 'loss' found in the denominator of the efficiency ratio over the vinculum to 'benefit' in the numerator simultaneously suggesting that a system with current x per cent efficiency should upgrade to future y per cent efficiency. This performance focus is mediated by high levels of complexity which means that 'freed up' paragains are unpredictable and lead to new unforeseen distributions of resources between interconnected users that might thwart prior expectations of who or what gets the salvaged loss.

The chapter introduced three types of liminal paracommons through which different types of resource systems transit when being pushed towards higher levels of productive performance. In each of these three types, though different, are arrays of pathways through which the principal resource and different fractions of loss flow. These uncertain and multiplicative 'liminal spaces' put social-ecological systems into a state of limbo wherein knowledge of specific efficiency-related consumptive, productive and distributive pathways and outcomes is fuzzy and incomplete. The paracommons can also be interpreted by asking who materially gains from an efficiency performance gain. In different ways, either because of the mental model constructed of the systems, or because of the nature of the system being studied, gains flow to the proprietor, its neighbour, the common pool or the wider economy. Furthermore attempts at raising efficiencies of large socio-ecological systems are often accompanied by prior expectations, setting up the risk of paradoxes.

Chapter 6

Distinctions between the commons and paracommons

By employing efficiency as the lens through which to view natural resource systems, the paracommons can be distinguished from the 'principal commons' (Table 23). In this chapter I have categorised these distinctions into two sections; a) resource conservation and b) subtractability and competition for gains. With respect to the paracommons, this division into two sections and their order is artificial because it is the second section on the competition for salvaged losses that partially 'solves' the first section on resource conservation by retaining or sending them to the common pool.

Prior to an explanation of some of those distinctions, I first address the term 'principal'. The principal commons describes a natural capital common pool of resources subject to human withdrawal and consumption. These are the commons that the common pool resource (CPR) literature is mostly concerned with (Baden and Noonan, 1998; Dolšak and Ostrom, 2003b). They exhibit problems of 'non-excludability' (difficult to exclude users); and 'subtractability' (or 'rivalry', where in joint use, one user is able to subtract welfare from another). Examples of rivalry exist over fish from a freshwater lake, or over water drawn from a stream.

Furthermore, I combine both stocks and flows as constitutive features of the 'principal commons'. Although distinctions in ecological or economic systems between stocks and flows can be made (see Millennium Ecosystems Services and related literature (MEA, 2005; Mooney et al., 2009)) where a stock of a resource gives rise to portions that can be harvested as a flow, the central idea of a principal commons is of a linear flow of resources with waste as an externality to the system of consumption. Thus both the body of water in a dam and the dam's exit flow comprise 'a principal commons' with irrigation drainage as one of its 'wastes'. Another 'principal commons' example includes the adult fish in a fishery plus the fish eggs and fry, with fish discards as a waste.

Table 23 Distinguishing the commons and paracommons

	Dimension	The (principal) commons	The (liminal) paracommons
Resource conservation	Conversion processes	Resource replenishment, recovery and regrowth. Sustainable harvest attempts to match resource replenishment.	Multiple conversion concepts: e.g. resource to wastage (harmful, recovered, not withdrawn, avoided, non-recovered).
	Efficiency and performance ratios	Given brief, incomplete or summary treatment. Perceived to be linear and straightforward.	Defining feature of the paracommons. Complex, obfuscating, nested and scalar.
	Withdrawal (harvest)	Withdrawal/appropriation.	Intrinsic and extrinsic withdrawal (wastage/wastes elements).
	Resource conservation	Reduction of abstraction/ withdrawal.	Reduction of intrinsic or extrinsic withdrawals.
	Regulatory questions and CPR principles	How to manage CP resources, and regulate demand. What CPR modes?	How to govern the commons with a significant proportion of salvageable wastage? What technologies and prices?
	Design and technology	Technology related to harvest and appropriation capacity.	Technology related to raising efficiency – though often unpredictably so and requiring 'systems thinking'
Subtractability and competition for losses	Subtractability	Subtractability: a resource consumed in one place cannot be consumed elsewhere.	Modified subtractability: a resource consumed in one place leads to products or resources captured/reused elsewhere.
	Non-excludability or rivalry	Defining feature of the principal commons. Difficult to exclude others accounting for geographical and spatial factors.	Excluding others from accessing the waste fraction is one option; or not if recycling/reuse by others deemed normal.
	Spatiality-conferred ownership claims and competition	In parallel (all users acting simultaneously); or in geographical sequence with users in longitudinal, vertical or lateral sequence.	Extrinsic-appropriation sequence: likely to be a complex intricate and unique set of resource and wastage flows in different contexts.
	Property rights questions	What goods require what property regimes? Who owns the commons? How might rights be transferred?	Who owns the waste and recycled resources, and future waste saved and averted? How is ownership of paragain ('tare' portion) transferred?
	Pollution, wastes and wastage	Summary treatment or seen as pollution subject to removal and treatment. Pollution becomes an externality.	Materials are desirable and subject to competition and recovery. Wastages and wastes may be non-local but are an internality within a wider system

.

6.1 Resource conservation and conversions

6.1.1 The defining conversion process and efficiency ratios

An interpretation of the conversion process at the heart of the sustainable harvest and protection of natural capital defines the two types of commons. Taking fisheries as an example, the conversion process at the centre of the principal commons is that of a natural biological renewal process, for example, viable adult fish spawning eggs and fry. To sustainably manage fisheries is to set harvest rates against the rate of reproduction to protect fish stocks so that they replenish themselves.[1] In the forestry commons, the conversion is one of trees seeding new trees. From seeking this balance stems the problems of entitlement to harvest associated with the principal commons.

However, in the paracommons, the conversion process adds elements of managing losses in the chain of husbandry, harvesting, processing, storage and production. Not only does this introduce another component (losses) to address sustainability and balance supply with societal demand, but the translocation of losses takes place to the same or neighbouring user, to the common pool or wider economy, all accompanied by expectations of who will get what salvaged losses. The conversion processes increases in complexity and distance, requiring carefully derived measures to judge performance, specifying how scale and different pathways are accounted for. Efficiency ratios, though appearing to be effective in capturing resource management performance, fail to fully describe the nuances of multiple scales, pathways and resolutions.

6.1.2 Withdrawal of resources from the common pool

I define two types of withdrawal (synonymous here with appropriation) – intrinsic and extrinsic. The principal commons witnesses water abstraction with little recognition of this distinction (Ostrom, 1990). Without this distinction, all withdrawal is deemed to be intrinsic to the user's needs and therefore the total withdrawal establishes concerns regarding the subtractable and rivalrous nature of the principal commons.

However, in the paracommons, both intrinsic and extrinsic withdrawals exist. The intrinsic element is the beneficial and consumed fraction (e.g. crop transpiration) while the extrinsic element is the additional fraction abstracted to meet expected recoverable and non-recoverable 'losses'. For example, a sizeable extrinsic fraction may be normatively engineered to be part of the intrinsic net crop water needs, observed in irrigation. Yet intrinsic but especially extrinsic withdrawals are not objectively defined, thereby creating political or scientific space for claims of waste and excess in the face of competition. This subjectivity, combined with re-use of part of the extrinsic withdrawal, disputes the nature of resource subtractability and rivalry, making these far more ambiguous (see also Section 6.2.1).

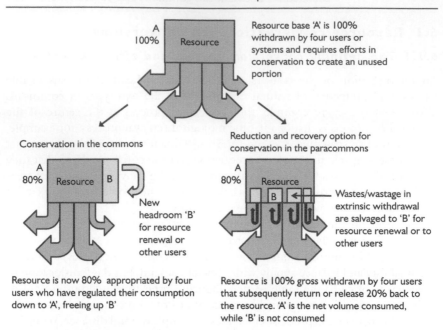

Figure 45   Distinguishing conservation in the principal and paracommons

6.1.3 Resource conservation in common pool resources

As the left-hand side of Figure 45 shows, conservation in the principal commons takes place via the reduction of aggregate appropriation. The conserved resource then plays its part in renewing the resource be it fish, grazing land or forest. Although conservation within CPR literature is discussed, this is largely related to the provenance of conservation thinking in ecology and the preservation of renewal in biological populations (e.g. fishstocks, see McCay, 1996). The attention afforded to a comprehensive view of physical resource efficiency – and the mechanisms to improve efficiency – tends to be neglected or at best underwhelming. For example, Ostrom *et al.* (1999) and Dietz *et al.* (2003) omit the complexities of governing the commons arising from resource inefficiencies.

In an increasingly competitive world, matters accelerate. Waste fulfils little beneficial function to some resource users, but to others or society as a whole, it may rapidly hold considerable value. The paracommons therefore offers conservation via a reduction in both intrinsic and extrinsic withdrawal. The extrinsic withdrawal gives rise to different options (e.g. Figure 38) for the conservation of a resource ordinarily deemed and expected to be wasteful. Furthermore, the paracommons recognises that for conservation to be a priority, the paragain from reducing the extrinsic portion is retained by or returned to the common pool. Thus the paracommons arises because of the number of choices to conserve a resource, and multiple political interests levied against the 'yet to be wasted' principal resource. These options for arriving at the aggregate consumption from the common pool of one abstractor, and the consequence impact on other abstractors also depend on accounting formulation. Furthermore, these formulations may not be widely agreed, creating real-world ramifications for how claims and counter-claims are constructed.

6.1.4 Common pool resource 'principles'

A great deal of work has been done on the original Ostrom design rules or principles that might guide the successful conservation and management of common pool resources in the face of potential unfettered demand. There is not the space here to review the refinements applied since the original principles were identified by Ostrom in her studies of irrigating water user association (see Cox *et al.*, 2010). Nevertheless by outlining the CPR rules/principles (Box 4), it is possible to see that 'efficiency as resource' is only partly captured. Particularly challenging to resource users are the boundary problems associated with losses moving from one level of scale into another level, plus coping with highly appropriative contexts (of the paragain), and the problems with attribution (who is responsible for losses in a chain of abstraction and use and when these arise through multiple sources).

The principles in Box 4 can be cross-examined in the light of the paracommons, asking in effect what principles apply to paragains (rather than principal resources) freed up by efficiency improvements. Reflecting on this, a few illustrative cautions are suggested:

Box 4 Principles for common pool management, after Ostrom (1990)

- Clearly defined boundaries of the system so users and their benefits more understood and predicted.
- Proportional equivalence of costs and benefits for each user.
- Collective choice arrangements (users of water generate own rules for belonging, employment, duties, punishment).
- Monitoring: users provide monitors, in turn are accountable.
- Graduated sanctions: violators exposed to sanctions, which are small in first instances.
- Conflict resolution mechanisms: impartial, low-cost, local body/ mechanisms to resolve conflicts.
- Minimal rights to organise: outside institutions will not challenge authority of local institution (+ formal recognition).
- Nested enterprises: multiple layers of organising groups.

- The delineation of boundaries to be drawn up requires particular care because losses move laterally to neighbours and also up levels of scale.
- The identification of the size or volume of current 'losses' carefully ajudged by stakeholders against possible future or more prudent scenarios.
- The question of distributing costs, benefits and sanctions of raising efficiency to users is largely unresolved. Some costs at the unit level will be borne by individuals, other costs by groups and others, at the basin scale, by donors and/or the state.
- The somewhat abstract idea of how to achieve and then steer a paragain – the material gain of an efficiency gain – requires deliberation by resource users. How 'losses foregone' arise in a recursive web of resource use are difficult to isolate. In a highly appropriative context, steering paragains to the 'common pool' to conserve environmental stocks may require special consideration.
- The conflict resolution and measurement mechanisms that surround efficiency and losses – especially where losses for one user might be a gain for another – are also undetermined.

Underpinning all of these is a lack of accurate quantitative metrics of resource distribution and performance. Because of this resource users and other actors generally do not know how the 'CPR principles' or 'paracommons principles' result in systems being more equitable or efficient beyond opinion, assumption and expectation. Putting efficiency more squarely into the CPR frame predetermines that some form of measurement and monitoring – not only of rules adherence – of efficiency and distribution is a prerequisite.

6.1.5 Addressing conservation by addressing prevailing assumptions

Picking up on a discussion from Chapter 4, the design of irrigation systems illustrates the manner in which 'losses' are artificially or axiomatically created at the outset which then need to be reduced by various means. The contrast made between assumptions applied to irrigation efficiency (how much water is extrinsic to net crop transpiration) and the ensuing physical, extant realities once the system is built and up and running, suggests it might be more appropriate to begin with a leaner set of assumptions so that conservation is built into the design and management of systems. Instead, losses are 'gifted' to the proprietor system as a part of design – unsurprisingly assisting an ensuing appropriation of any paragains.

Irrigation engineers employ highly standardised procedures (FAO, 1977, 1999) to establish the dimensions of irrigation to meet crop water requirement and to design headworks as abstraction points on rivers and the canals that feed irrigated fields. This procedure includes five variables: crop water requirements; a dependability measure (usually to meet four in five years of rainfall); command area; supply time (usually 24 hours); and a measure of irrigation efficiency. This is procedure well established throughout the world when trained engineers become involved in drawing up water demands for irrigation systems.

While a number of uncertainties exist with each component (for example, modelling crop water requirements), the officialised procedure offers opportunities for accessing excessive water thereby depleting downstream users of their water (Lankford, 2004) and for poorly accommodating the differences between design and construction assumptions and eventual outcomes. It does this in six interconnected ways:

- First, irrigation efficiency is rarely measured. This is because irrigation efficiency is extremely difficult to assess but also engineers habitually look up or estimate figures rather insist on measurements (difficult if no scheme has been built).
- Related and second, the assumed design efficiencies are commonly too low. For example, in Tanzania the rice water duty accepted by government engineers is 2.0 litres/second/hectare (l/sec/ha) largely via the low efficiencies applied to system design. This contrasts with a measurable demand of about 1.0 l/sec/ha or less (Machibya, 2003). This establishes systems that abstract more water than if tighter parameters had been used (see also Lankford and Gowing, 1996a, for similar findings in Malaysia and also Murray-Rust and Snellen (1993) on design approximations affecting performance).
- Third, designers and regulators incorrectly assume that water wastage will be returned to drains despite commonplace evidence that farmers rarely care for their drainage water and that uses evolve around such spillages (be they opportunistic farmers, sinks or wetlands).
- Fourth, current design procedures prioritise the irrigation system rather than split water between irrigation and downstream (Lankford, 2004). Put simply, the design standard has no formal step for taking potential downstream demands into account when adjusting for improvements in operational or design efficiency and often tends towards designs that are non-proportional in their division between irrigation flows and river flows.
- Fifth, differences between design and final reality are rarely purposively revisited in the light of errors uncovered. While farmers might make adjustments either via negotiated or conflictual means (e.g. vandalism), often these tend to be *ad hoc* (replacing steel with wooden gates or adding sandbags), rather than substantive (replacing an undershot orifice gate with a proportional gate or installing a larger or smaller discharge orifice).
- Perhaps most significantly, sixth, the design parameters and physical works are often underpinned by a legal water right to withdraw a given amount of water. This means that 'losses' are both bundled into a water right and physically within the concrete dimensions of the canal headworks (further discussed in Section 6.2.4, below).

These procedural 'regularities' establish resource capture irregularities when scarcity and multiple demands come into play in a situation that differs from the original (often erroneous) assumptions about efficiency. The consequences for water consumption and allocation following planned or unplanned improvements in irrigation efficiency are profound because the design and its operation biases the system to utilise those 'savings' within the irrigation system rather than cascade them up to the headworks for downstream users (via avoided withdrawal or reduced forestalled fraction). In other words, the six points above form highly appropriative conditions in favour of the proprietor system. Even in the latest monographs on irrigation design (Bos *et al.*, 2009) the above flaws are poorly covered. Smarter, leaner assumptions for irrigation planning might help reduce differences between protocols and outcomes; by designing in scarcity, the proprietor system works harder to accommodate additional scarcity (e.g. a drought) by other means such as retrenchment and dematerialisation.

6.2 Subtractability and competition for paragains

6.2.1 Subtractability and rivalry

I question the purity of the concept 'subtractability' as a defining feature of the commons. Subtractability is explained in a discussion on resource units by Ostrom (1990, p.31) *'Resource units are not subject to joint use'* (her italics) and repeated by Dolšak and Ostrom (2003a): 'The tons of fish or acre-feet of water withdrawn from a particular water resources by one user are no longer available to others using the same resource' (p.7). In the principal commons, a unit consumed by one user is no longer available for another. For example, fish stocks are 'subtractable' when fish discards at sea benefit only the marine food-chain.

In the paracommons, the loss portion of a unit withdrawn by one user is available for other users including the original, proprietor withdrawer through different pathways. Moreover, this extrinsic portion is commonly modified by location, quality, quantity and timing – influencing the manner in which ownership claims are mobilised, pursued and countered. Therefore the paracommons features 'modified subtractability', where a loss becomes available for other users. This happens in two main ways. First the wastage/waste is generated by one system and recovered by another, for example, where a farmer picks up drainage water from an upstream irrigation system. Second, the loss is forestalled (or avoided) by one system for use elsewhere in the river basin. One might argue that the latter is not a case of reuse and does not circumvent subtractability. While technically true, in fact the highly appropriative environment means that this fraction is likely to redeploy to the proprietor system to extend its command area. But this paragain is not intrinsic to the proprietor's needs and is in the face of expectations by other users, particularly neighbours. While it is used by the proprietor in their inefficient system today, it need not be used in their more efficient system tomorrow. So while it is consumed and used physically, there are potential 'shadow' uses and expectations which competitors must wrestle with.

6.2.2 User/system interconnectivity

A source of complexity in resource efficiency use arises from the interconnections between resource users caused by both (in)efficiencies and the destination of any material gain freed up by an efficiency change. This means that resource users already facing rivalry over intrinsic principal resources are in competition over the hidden inefficient 'tare' fraction and the paragain. As Chapter 4 pointed out, these interconnections and competitions take place within levels between proprietors and neighbours, and across levels between proprietors and nested systems. These interconnections have ramifications for the spatial nature of competition, a point picked up next.

6.2.3 Spatiality-conferred claims and competition

Spatially determined access to the principal commons such as fish and wildlife are influenced by their population distributions, rights-defined territory, or by the physical location of the abstractor (with additional difficulties mediated by power relations, controlling regulation and excluding access). Boundaries have relevance here (for example, Acheson and Brewer (2003) discuss physical territorial boundaries) which a fisher or forester (or other harvester) may not exceed or transgress. Furthermore, surface water, in its principal commons form cascades through the landscape longitudinally, vertically and laterally, accentuating social power relations, exemplified by top-to-tail-enders in irrigation systems and river catchments (Zeitoun and Warner, 2006).

However, additional to these spatial patterns, the paracommons contains another spatiality – that of a sequence of abstraction-consumption-loss-recovery. Thus, the spatiality of losses to be recovered and salvaged is introduced. This will be complicated and particular to each system and context. If irrigation is an example, the location of fields, canals, pipes, drains, owners, neighbours, licences and rights create a special mosaic of waste/wastage sources and of legal and unsanctioned recovery. But significantly this sets up a current pattern of reuse-mediated spatial ownership. It is the reworking of a system to *raise* efficiency that changes the physical pathways that water then follows, and whether this takes place quantitatively, for example, where low-salinity drainage water is reused) or qualitatively (saline drainage water generated). Figure 39, above, shows that reworking the efficiency of an irrigation system will change the spatial patterns of current water fractions into something new.

Moreover, it is equally important to recall, this 'before and after' is mediated by *assumptions* of what is currently taking place and *expectations* of what will happen to the *potential pathways* that wastage might follow (or are foregone into different fractions) before they physically finally flow to different destinations within the landscape, thereby decomposing back into the principal commons.

6.2.4 Property rights and water legislation for water conservation

Perhaps the most exciting and relevant way of demonstrating the paracommons and of 'efficiency as resource' can be traced to deliberations on water legislation in the dry parts of the United States of America (Pease, 2012; Norris, 2011) and Australia (Crase *et al.*, 2009). These cases exemplify the problem of accommodating the revised destinations and ownerships of the volumes of 'new' water made available through water conservation (Skaggs *et al.*, 2011). In so doing these accommodations reveal the uncertainties of the paracommons, current structures to deal with recoverable or avoidable fractions and the extent to which existing water law (with reference to 'prior appropriation in Western US states) acts in the appropriative interests of actors and states (Davenport, 2008).

Bell (2007) describes attempts in four US states to remove disincentives to water conservation which existed under a previous prior appropriation doctrine (a problem earlier highlighted by Huffaker *et al.*, 2000). The previous doctrine forbade the spreading of the freed up or salvaged water to new lands belonging to the water right holder as this effectively expanded their water right expressed in consumptive terms. The four examples show who then had claims to the 'freed up' water including the original user, other users/sectors, the state for further allocation, and freshwater ecology. Identifying uncertainties for the programmes, Bell notes that a lack of take up in Oregon can be related to the high costs of conservation to meet aims, and identifies the differences between expectations of benefits and ensuing reality, writing on page 3:

> While Montana's salvage statute provides the opportunity to better use limited water resources, determining whether the conservation measures implemented actually save water can be difficult and complex. These difficulties have limited the success of Montana's program. The Department of Natural Resources and Conservation has noted that permitting an applicant to enlarge their irrigated acreage based on the water saved when switching from flood irrigation to a sprinkler system may diminish return flows, thereby injuring junior appropriators or other third parties.

Neuman (1998) and Shupe (1982) provide further analysis of the manner in which entrenched water abstraction law (and water markets) in the western United States 'has revealed itself to be woefully inadequate at eliminating waste and encouraging efficiency. Beneficial use affirmatively protects inefficient water use customs and practices' (Neuman, 1998, p.996). Included in the wider structures of legislative change are compensations arising over 'takings' where the state on behalf of itself or other parties elicits public property rights by taking up existing private property rights. Agreeing with Cole that 'takings' are a boundary issue (2002, p.166), in the case of conservation, the paracommons, with its multiple pathways, inexactitudes and opportunity for claim-making, constitutes a thicker form of boundary. As society witnesses new levels of resource competition, connections and fragmentation, we necessarily require new regimes for legislating for the intrinsic and extrinsic components of resource consumption separately and for meeting the transaction costs of doing so.

In the 'Tragedy of the Commons' (1968), Hardin believed that the individual obtains direct personal benefit from abstraction and consumption, with the costs of this borne collectively by others. He argued that profit maximisation drives forward the over-use of the resource, and that conservation simply frees up the resource for others to use. On the other hand, a rather conventional reading of efficiency is that the 'freed-up' resource will be used by the proprietor. However, by taking a paracommons view, the 'tare' portion bundled within extrinsic appropriation, achieves more visibility in policy and science, albeit a complex one. Moreover this portion can be reduced and isolated to flow to one or more of four destinations, under control or not. That said, whether new recycling and paragain property rights can be forged that both reflect what is possible on the ground and drive new sustainable and equitable patterns of consumption will arguably be a key contestation of the paracommons.

6.2.5 Pollution in common pool resources

As wastes are the central focus of the paracommons, the treatment of pollution in common pool literature is relevant. Cole's (2002) analysis is instructive in so much as while he refers to pollution as a defining feature of the commons – that pollution is largely an unwanted by-product of the absence, or inappropriate mix, of property rights and control mechanisms with attendant excludability, informational and transactional failures. While I have no argument with this per se, this analysis applies toxicity or nuisance values to pollution in keeping with Hardin's view that the gains of resource use fall to a few, but costs of pollution fall to all (or to the poorest). This is the same starting point for (in) justices analysed by Low and Gleeson (1998) and Walker (2012) that, albeit resource and waste distribution is treated as amongst the factors that determine distribution of environmental justices, waste is seen as without value or with negative harmful properties, and otherwise an injustice. Despite Dolšak and Ostrom (2003a, p.15) drawing our attention to the problems of devising rules for a sink-type commons (where air soaks up gases or particulate pollution), this analysis does not draw attention to potential routing pathways of recyclable and recoverable losses over and through transitive options spaces, and their respective characteristics, costs and benefits. In other words, the commons literature does not explicitly frame pollution as a desirable common resource. It is when pollution or wastes become valuable, and changing as a result of efficiency interventions, we have the option of new property rights overlapping or of one resource leveraging another.

6.3 Chapter review

Using two themes of conservation and competition, a number of key comparisons between the 'principal' commons and the paracommons were explored, chief of which is that rivalry and subtractability do not apply to the latter because loss fractions can be re-used or forestalled/reduced and re-deployed. This distinction allows us to distinguish other differences such as how pollution, conservation and property rights are treated. It should also be recalled that a recycled/forestalled resource can physically and socially influence the principal resource because of the assumptions wrapped up with the latter, often lending the proprietor system a variety of ways to appropriate any salvaged losses.

Chapter 7

Significances and applications of the paracommons

7.1 Policy prefiguration and the politics of efficiency

These two sub-sections contain further analyses of the two policy-related points on liminality introduced in Section 5.1.

7.1.1 Policy prefiguration and paradox

Formal well-intentioned policy-driven attempts to raise the efficiency of a large system or sector are central to the concept of the liminal paracommons. Policy intends to raise performance (setting up the chance for releasing paragains) but tends to be inadequate in understanding efficiency complexity with the consequence that outcomes differ from original intentions – disappointingly or paradoxically so (see Lopez-Gunn *et al.*, 2012). This inadequacy arises because of the accelerating degree of complexity moving from small individual conversion units such as fields and plots (or indeed light bulbs) to agglomerations and collectives such as irrigation systems, cities or whole irrigation (or energy) sectors. This turns efficiency from being a relatively simple and controllable endeavour in the first instance to a matter of difficulty and uncertainty in the second instance. A comprehension lag arises from a policy-maker's mistaken inference that large systems are simply bigger versions of smaller systems, rather than allowing for emerging behaviours that come with larger systems.[1]

In terms of policy, while the concept of the paracommons describes natural resource systems with prospects that could fall to different resolutions, it is not merely the 'before and after' of an intervention applied to natural resources. Rather the liminality of the paracommons pertains to the differences between the 'assumed yet in reality unknown before' and 'expected yet unknown after'. The mistaken views on 'before' (or today) arises because of current inadequate measurement of resource inputs, outputs and ratios and the excessive dominance of prevailing yet outmoded efficiency protocols, narratives and solutions. The expected 'after' arises out of the naïve assumptions of the benefits of future efficiency/productivity changes and their effect on total resource depletion rates and distributions to different users as well as continuing gaps in measurement. The paracommons occurs because both prefigurations are uncertain. Furthermore liminality should not be applied to, say, changes within a 'principal' commons such as a sea fisheries where these do not witness parallel, valuable and sizeable waste streams and complex mechanisms and pathways for recovering, economising or avoiding. Given the multiplicative nature of the liminal paracommons wherein many outcomes are possible, conditions favour proprietor appropriation of gains and measurement is sorely lacking, paradoxes are likely. The idea of 'prefiguration' covers this sense of mistaken 'representation beforehand' well.

7.1.2 The politics of resource use efficiency

However, by purposively naturalising efficiency and productivity (Boelens and Vos, 2012; Clement, 2013), can the politics and policies of resource use efficiency be construed as maladaptive as well as naïve (the point above)? In other words, might efficiency-predicated policies lead resources and users in directions that result in greater imbalances in equity and access, to lower efficiency and higher consumption of natural resources by certain favoured actors? There is insufficient space (and empirical data) to examine this question in detail but one may ask whether the term 'paradox' employed in efficiency analyses suggests that policy may too often be wilfully inappropriate (and appropriative) as well as innocently naïve. Certainly the risk is ever-present, as Clement (2013) infers with reference to water productivity concepts:

> water productivity discourses represent inputs and outputs as resources naturally flowing into and outside of the system but have hardly considered issues of access and control over the resources or their benefits. This supposed neutrality neatly conceals the politics of water management.

The marginalisation of resource users as inefficient producers by governments following various modernising agendas (van Halsema and Vincent, 2012; Lopez-Gunn *et al.*, 2012; Boelens and Vos, 2012) is interpretable in different ways using a paracommons framing of efficiency and productivity. *Accurately* rendering water users as inefficient may, on occasion, be narrowly technically appropriate as long as reuse of water within a nested system is expressly not included, or where a wider political and market economy is transparently acknowledged for discouraging farmer investments in infrastructural and institutional innovation or where the landscape and mosaic of water use is objectively uncomplicated.

However, the paracommons tells us that dangers lie with political and commercial strategies to gain control and influence over resources managed by poor communities on the basis of narrow efficiency reasoning. Terms such as 'losses', 'waste' and 'efficiency' buttress truth-claims (Boelen and Vos, 2012) for ensuing modernisation programmes to raise efficiency and to gain control over resources (such as converting small farms with earthen furrows into larger canalised systems). As Boelens and Vos acutely observe, these programmes explicate government intentions towards environmental stewardship through the objectification of efficiency and productivity with little real understanding of their composition and conception, or of how farmers come to understand them. The outcome, also recorded by Boelens and Vos and other authors (Lopez-Gunn *et al.*, 2012; Knox *et al.*, 2012; Trawick, 2001) is an inability to partner with farmers to solve either national-level concerns regarding food production or local-level concerns regarding top-tail end water sharing.

Seemingly sensible interventions (e.g. canal lining, adopting treadle pumps, promoting pressurised and drip systems, pricing water) promoted in the name of efficiency may in many situations be wholly inappropriate depending on both wider basin circumstances and local contexts and trajectories. Critical grounded policy that has a central focus of changing practice and technology over thousands of hectares of surface irrigation should be seen as very different from trialling fashionable ideas at the periphery (e.g. bucket drip kits or irrigated home gardens in Sub-Saharan Africa). Thus, while micro-irrigation projects promulgated by international and national institutions and NGOs may hold an initial attraction, they should be seen as political projects first and foremost and second, as startlingly naïve in terms of effective ways of irrigating at scale, and for their ability to free up losses for use elsewhere. As mentioned above, some technologies should be viewed as being appropriative, allowing efficiency gains to be immediately recaptured locally.

The paracommons provides ways to soberly assess regulatory and control regimes favouring 'efficient' modern technologies. For example, new policy might be preceded and accompanied by validated, fully costed, aggregated evidence of current and future patterns of water depletion and distribution. Also, one might seek clarification on the objectives of an efficiency programme (proprietor performance, neighbourly and environmental equity and distribution to other parties in the wider economy) with the understanding that meeting all four transparently within the parasystem given highly skewed appropriative drivers is unlikely. It is also reasonable to suspect that larger more-efficient systems of production will subsequently drive up the use of other resources such as energy.

Furthermore, how policy establishes, convincingly, the methodologies, resources, skills and services to achieve real improvements to efficiency is a subject of profound importance. How ironic is the contrast between government accusations of inefficient resource users and the equivocal government structures and services in place to achieve higher efficiency and to transparently 'bank' material gains – paragains – to a chosen destination. As currently construed, techno-centric solutions such as drip, canal lining or farmer training are unlikely to match the complexities outlined in REC and the paracommons. Although other sections in this book hint at a much more cautious and comprehensive approach to managing efficiency and productivity upwards – it is not the function of this book to specify a REC log-frame for such an approach. Nevertheless, a key methodology would be to probe the mix of understandings held by resources users and the managerial professions. An acknowledgement that irrigation efficiency and productivity are complex by-products emerging from multiple field-level factors managed by farmers would dramatically change the relationship between irrigators, ministry engineers and other state actors.

These issues and questions characterise a political ecology of irrigation efficiency (Lankford, 2012a). They feed a critique of the framings and narratives of environmental governance and management (in this case irrigation water) through the co-production of actors, knowledge, theory, policy and practice. Irrigation efficiency is an ideal candidate for environmental and agricultural myth-making. Far from being static, it has reflected the shifting and multiple emphases placed on irrigation as an agronomic input, engineering activity, socio-technical food production system and major consumer of freshwater (see Table 7). The classical efficiency form should be dissected for its inability to underscore water allocation. Furthermore, CIE has been misemployed politically for the purpose of soliciting support for policy and fiscal expenditure to deal with falsely low efficiencies. Efficiency is well placed for new discourse deconstructions and as an illuminating prism through which to gauge 'science', in its discursive, economic and environmental guises, struggling with scale, fluidity, variation, technology, people, institutions and fashions.

7.1.3 Enlarging policy space: irrigation productivity and water allocation

The paracommons invokes an idea of a larger more cautious policy space regarding interventions for conservation, productivity and allocation of gains to other destinations. This idea is explored in river basins with substantial areas of irrigation, where irrigation efficiency potentially plays an integral role in water allocation. Policy theory can be constructed too easily: in some river basins, the 'spare water' for other needy and pressing sectors is to be found within irrigation inefficiencies – within the water losses. The script reads 'reduce water demand for irrigation without impacting upon net needs for crops and spare water is freed for allocation'. Referring to Figure 38, this is pathway is 'forestalled fraction' (FF, or route numbered '12').

However, this pathway is uncertain because of other depletive and appropriative factors such as design and property rights. Thus a number of authors believe water conservation efforts raise consumption and reduce water for other sectors by invoking two other (mutually exclusive) pathways. First, there is widespread belief that 'lost' water from irrigation systems is not lost (see Keller *et al.*, 1996) and that it is recovered by the river basin by other sectors or irrigators. This is pathway number 11 in Figure 38. Second, the 'opposite' pathway is number 5; increasing efficiency leads to increased consumption by the proprietor which reduces return flows to aquifers and drains. Both routes are linked by Ward and Pulido-Velázquez (2008), who argue on page 18216:

> So, irrigation technologies that apply water at optimal times and locations in plant root zones increase crop consumptive use of water and crop yield as irrigation efficiency increases. When yield goes up, ET typically rises. Water losses through deep percolation or surface runoff will be reduced, possibly to nearly 0, through drip technology, but more ET will be used by the plant in supporting its reduced plant stress and higher yield. More efficient irrigation systems reduce diversions from streams and increase crop both yield and gross revenue (18). Depending on the cost of installing drip irrigation, costs and returns of production, and the price of water, the farmer who uses the technology may experience increased yield and higher income per unit of land. From the farmer's economic view the new water-conserving technology is good. However, basin-level consumptive use of water can increase.

While I have considerable sympathy with the caution expressed here, I hold the view that invoking 'only' these three pathways (5, 11, 12) fails to recognise the uncertainties as well as options at the heart of the liminal commons. First they are generally based on modelled assumptions rather than long-term trans-scalar fieldwork (from lysimeters up to basin scale hydrometry); second they ignore other available pathways; third they assume no regulation of withdrawals at the headworks; and fourth they assume water re-routed via aquifers and the environment is costless. With regards to the latter point, water 'lost and recaptured' rather than 'forestalled (or avoided) and released' (and therefore retained in fast flowing rivers and channels) ignores significant water productivity costs such as salinity and quality problems, location uncertainties, opportunities for further evaporation, and timing delays of water scheduling – the latter for living crops and animals dependent on water (Lankford, 2006, 2011). Put simply, the space for policy is increased by employing the 16 pathways of Figure 38.

7.1.4 Socialising efficiency

Earlier I wrote of the importance of seeing efficiency through the eyes of those closest to the consequences of inefficiency.[2] These might be users facing water scarcity or at the receiving end of efficiency governmentalities (following Agrawal, 2005). Moreover, users might be debating with their upstream neighbours whom they perceive as being 'wasteful' which, in a bifurcating model of resource provision, harms their production. Understanding and listening to these viewpoints harnesses a more socialised view of efficiency. There is another reason. Despite the accounting, scale and technologies of efficiency which place certain basin-level responsibilities on policy-makers and scientists, efficiency is promulgated at the field and household level by resource users. A river basin is a collection of, say, tens of irrigation abstraction points, hundreds of canals, thousands of farmers and tens of thousands of small plots and furrows. Thus there is a balance between interventions that reduce water withdrawal at the river basin scale and interventions that reduce water demand at the field scale. Central to both will be the inclusion of farmers in participatory agreements to deliver localised efficiency improvements (Lankford, 2006).

I draw attention to a participatory model I devised to explore local viewpoints on efficiency. The river basin game (Figure 46) reveals water knowledge held by farmers originating from different parts of irrigation systems or river catchments (Lankford and Watson, 2007; Rajabu, 2007). The game uses marbles and a sloping board to simulate water flow in a catchment and to mimic the tendency of top-enders to access more water than those located at the bottom of the board. Thus the game encourages resource users to behave as if they are top- or tail-enders. The left-hand photo shows resource users gather around the game with the facilitator at the top end. The right-hand photo of Figure 46 shows information gathered about local understandings of careful and wasteful use of water once farmers and other water users are given the opportunity and confidence to share information in a playful yet structured format.[3]

Both game role-playing and user knowledge build on and are reflections of practices shown in Figures 47 and 48 taken on Kapunga Irrigation Project in Southern Tanzania. These contrast irrigation practices between rice producers in an upstream water-rich location versus those growing rice in a water-short downstream location. Both sets of photos are taken about five kilometres apart. Figure 47 shows soil cultivation (the top-ender uses a lot of water to soften soil while the tail-ender hoes in dry soil) and Figure 48 shows rice nursery cultivation (the top-ender has a wide boundary of open water and weeds, while the tail-ender carefully controls water through the use of a straw mulch and soil bunds).

But socialising efficiency is not only about gleaning opinions on efficiency and wastefulness. I consider the defining test of efficiency thinking to be how resource users and supporting engineers, scientists and policy-makers respond to inefficiencies. In my view, all stakeholders appear too ready to jump to fashionable or well-known solutions without according significance to solutions already put in place by resource users. In other words, when confronted with visible or voiced concerns about water wastefulness, the consensus inexorably moves to solutions such as pricing water, lining canals, micro-irrigation, drought resistant crops and so on. Rarely do workshop participants walk fields together or spend effort detecting the quiet and nearly hidden ways of managing water more efficiently. Observations of the latter in Tanzania include: building soil bunds more robustly; agreeing that farmers should be present when water arrives at their plot; levying a local village tax on land prepared for planting (land area is easier to measure than stream flow); adding in small quaternary channels some few centimetres deep to ferry water through fields more quickly also reducing spreading of water leading to evaporation; deciding to keep dry

(a) Playing the marbles game   (b) Discussions after the marbles game

Figure 46 Using a marbles game to explore local understandings of efficiency
Photos show (a) the river basin game being tested by farmers in Igurusi village, Tanzania
(b) after the game, farmers sharing their knowledge on how they manage water, carefully
or not

(a) At the top end of the irrigation system   (b) At the tail end of the irrigation system

Figure 47 Comparing land preparation in Tanzania – top-end and tail-end irrigators
Photos show (a) water being used by upstream farmers to soften soil prior to cultivation
(b) tail-end cropper using hoes to till dry soil prior to arrival of water

(a) At the top end of the irrigation system   (b) At the tail end of the irrigation system

Figure 48 Comparing rice nurseries in Tanzania – top-end and tail-end irrigators
Photos show (a) top-end rice nursery surrounded by large area of open water evaporation
(b) tail-end rice nursery covered in a straw mulch to reduce evaporation and little excess
water at the nursery edge. Machibya (2003) found that farmers in the area of photo 'b'
grew 3–5 tonnes rice/ha with about 925 mm consumed water per season while farmers in
the area of photo (a) grew about 2–3 tonnes/ha rice consuming about 1400 mm per season

season crops close to the canal headworks on the river to reduce conveyance losses; gathering to clean and weed canals, and; foregoing a deep wetting and soaking of the soil prior to transplanting.

But even farmers may not see these as being valuable solutions; faced with the prospect of workshops, new outside ideas and finance (and well versed in such modalities), many resource users readily echo expert opinion and 'high' engineering.[4] My whimsical solution to socialising efficiency is to become a waterist or watereer; the designation of a community scientist exploring the democratisation of expertise (Woodhouse and Nieusma, 2001) (see Box 5).

Box 5 On being a 'waterist'/'watereer'

Through education and employment, we fall into categories: social scientist, economist, hydrologist, crop biologist, engineer and agronomist. Such specialists tend to see the world in a particular way. Claiming to be interdisciplinary and good at running participatory workshops, when asked for solutions to address low-yielding resources, people with well-developed career specialisms usually fall back on their training and 'office'. For example, 'price water' says the economist; 'breed higher yielding crops' says the biologist; 'line irrigation canals' intones the engineer; 'introduce new water laws' proposes the lawyer; 'form user groups' argues the social scientist, and 'partner with drip irrigation companies' suggests the policy-maker interested in public-private partnerships.

But what would you do if you were a 'waterist'? A waterist is someone whose discipline and position does not come to the fore when asked for a solution. He or she sees the solution by looking at the resource via the eyes of a certain kind of resource user. A waterist seeks out certain kinds of irrigators in an irrigation system who: a) rarely have any disciplinary training, and b) are farming on boundaries and margins of resource availability. An example might be a farmer at the tail end of an irrigation system or cropping a particularly shallow and infertile soil. A waterist realises these types of farmers are least likely to represent themselves loudly at a resource workshop where the momentum is building towards expert or bold solutions, but if engaged with might contribute their experience on managing a resource prudently and carefully. A waterist might also see they have already instigated, with neighbouring farmers, various agreements to deal with scarcity – for example, rotating flows; cleaning canals; not planting after a given date; adding in extra canals where necessary; choosing local shorter-season varieties; and undertaking dry field preparation. A waterist (or watereer) sees these 'at-the-margin' irrigators as waterists, and becomes their champion.[5]

7.2 Towards cyclical and industrialising ecosystems

A key premise of the paracommons is the presence of paragain redistribution, in other words that dynamic and redistributive connections via salvaging losses are central and defining features. The paracommons arises because material 'gains' achieved by one user at one scale need to be validated by reference to other users, locations and scales. This also takes place as society seeks to make socio-ecological systems higher performing and to understand that endeavour in more calculative ways. Taking this viewpoint, I believe it is possible to interpret ecosystem services in two ways: linear and cyclical. The linear interpretation views services flowing from nature to a destination or system of use (which I believe appears to be underlying message of the Millennium Ecosystem Assessment (see MEA, 2005, Figures A and B)). Yet, in an increasingly scarce and populous world, resources are recovered, forestalled and competed over in the shadow of efficiency-driven redistribution. A more cyclical industrialised model may be more appropriate. Yet it is my view a linear ecosystem services interpretation, while acknowledging complexity (Ruth *et al.*, 2011; Norgaard, 2010), does not fully speak to the reciprocity between realms and redistributions of resources, losses, wastes and wastages.

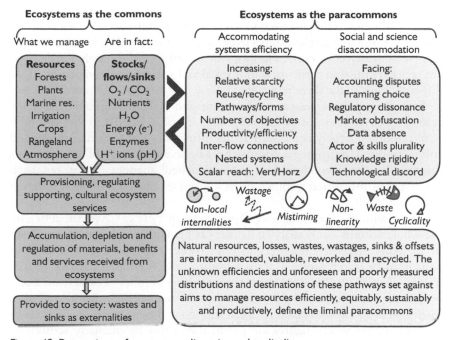

Figure 49 Perceptions of ecosystems linearity and cyclicality

This (somewhat artificial) dual interpretation can also be seen through the management of ecosystem services. I refer to the manner in which we utilise only a few instruments to address the multiservice benefits of nature – for example, land use planning and payment for ecosystems services (PES) to alter the rates of carbon sequestration, biodiversity and catchment hydrology. In other words, I cannot see how land use planning and PES by firmly planting us in 'ecosystems-as-providers' gives us more productive, efficient and reciprocal eco-agro-industrial-urban systems where the down-played denominator ecosystem services *should* be managed alongside the prioritised numerator ecosystem services. If system productivity and distribution is highly uncertain, complex and liminal, then payments alone to manage ecosystem services are unlikely to be effective. Many more managerial inputs and competences may be needed if society seeks to direct inputs to outputs with greater control over losses, wastes and wastages. A linear ES/PES venture may not solve the metabolic rift between nature and society (Clark and York, 2005).

7.3 Efficiency kinetics – resource management measurement

The commons and paracommons are connected to performance measurement and monitoring in four ways. These points are developed below, allowing then a brief discussion of a 'kinetics' framework of resource accounting to fill identified gaps.

My first point, recapping Section 4.6.6, argues that the accurate tracking and tracing of flows through landscapes over different times and scales is immensely difficult. Furthermore, current facilities for recording water supply and demand are generally missing particularly in tropical and sub-tropical river basins. This difficulty leads to the three points that follow.

Second, a lack of resource accounting and field measurement to track resource conversions, losses and recycling, arguably has led to the 'common pool resources' largely being seen as the 'principal' type featuring rivalry and subtractability. Controversial this may be, I believe that much of the commons research has been conducted in a 'decalculated' manner; it has not been interested in measuring resource flows and losses cascading though different scales and levels.

Third, although researchers work on productivity (e.g. CAWMA, 2007), on the whole water is rarely analysed by users as benefits to cubic metres depleted or in terms of related resources in ratio (water depleted per kilowatt hour consumed or water depleted per net tonnes carbon emitted). A lack of data means scientists and resource users cannot fulfil a number of objectives: assess output performance; audit the performance of management before and after institutional and technological reforms; and test and hone theories of efficiency accounting. Yet where figures are given, they must be intensely scrutinised because of the logistical and methodological challenges in measuring resource inputs and outputs within the appropriate boundaries and scales (van der Kooij *et al.*, 2013).

Fourth, one must question the extent to which existing water accounts are a reflection of a desire only to account for inputs and outcomes at the higher basin scale. While water accounting methods (e.g. Perry, 2007; Frederiksen and Allen, 2011) fulfil an important function in understanding how water is disposed of in river basins, it is legitimate to ask whether these methods help with the management of water resources in river basins. Four questions throw light on current approaches to water accounts and fractions:

- How are water accounts derived from measures 'in reality' (empirically at the unit, system and parasystem levels)?
- Can water 'outcome/disposition' accounts be utilised by irrigation managers seeking to raise efficiency of their proprietor system while capping total consumption?
- Can water accounts be used to track paragains to one or more destinations within parasystems?
- How can water accounts be utilised by policy makers to support sectors in becoming more efficient?

I am particularly interested in these questions because they relate to the details of water flows in irrigation systems and throughout the parasystem (users, neighbours, economy and environment). It is my concern that accounting 'only' for basin outcomes cannot contribute to more complicated decisions regarding technology and institutions that shape what takes place in many fields and farms each hour and day of the year.

Reflecting on the above four questions, I believe that a future field of theory and practice of resource management, characterised by increased competition over resources will necessitate more high-density, frequent and numerate resource measurement (Ostendorf, 2011). This will have legal ramifications because, I argue, that the currently inadequate assignment of property rights to water resources in relation to who owns the consequences of changes in productivity and efficiency can be traced to poor monitoring and evaluation (see also Skaggs *et al.*, 2011).

In response to these calculative concerns, I believe that the accounting and measuring of resource use efficiency could be captured under the term 'efficiency kinetics'. The word 'kinetics' covers 'those aspects of a particular process that relate to the rate at which it occurs; the details of the way a process occurs, especially as regards its rate' (OED, 2012). Kinetics goes beyond accounting for outcomes – it is interested in explaining resources and losses passing through myriad pathways to final outcomes and dispositions. Kinetics sits within the liminal space between inputs and outputs and denotes a much more comprehensive field study of resource use, efficiency, productivity and distributions than at present. In the spirit of rigorous accounting (Perry, 2011) but also open to evolution and local user views and metrics (Trawick, 2001), resource efficiency kinetics would attend to a number of objectives. I outline the following principles of a more engaged approach to resource efficiency kinetics:

- Explicitly recognise the problem of scalar levels in assessing the quantities and pathways of resources and their flows. This means putting measuring stations at different levels (field, stream, basin) to be able to collate a full picture of how resources move.
- Be action-oriented towards the evaluations of technologies and assemblages of technologies aiming to raise performance. This means experimenting with new ideas to see how they affect resource performance.
- Incorporate water users and other actors immersed in, and affected by, resource efficiency. Because cascading resources exhibit neighbourliness when divided and bifurcated, reference must be made to the views of those affected by their neighbouring users and systems.
- Continuously refine terminologies and definitions for evaluating and communicating efficiency and productivity. The language and communication of the paracommons are integral parts of how different stakeholders with very different views come to accept what is happening within and around their systems. Nonetheless, dual terminologies are likely.
- Involve multiple methods to cross-check the quantities, qualities, locations and timings of resource flows. Irrigation efficiency cannot only rely on measuring canal and field losses and multiplying them together.

Efficiency kinetics might be constructed from three 'accounts' of measurement and ongoing monitoring (Figure 50 and Table 24). The three accounting approaches have their equivalent in financial accounting for companies and corporations. Final disposition water accounting is similar to 'final accounts' or year-end profit and loss accounts. These accounts are can also be termed statutory accounts. Then there are on-going management accounts which are generated on a regular basis (usually monthly) by company managers to monitor the running of their company. Lastly there are 'appropriation accounts of profit and loss' which are used to track how profits are distributed to different parts of an organisation. In complex resources exemplified by irrigation, these do not form easy categories; they cross-refer and form intermediate or hybrid versions. The latter two are discussed next.

Management and appropriation accounting

There is insufficient space to spell out a detailed methodology that probes irrigation efficiency kinetics, particularly as the three water accounting types in Table 24 will have several methods. With this in mind, I take a page or so to

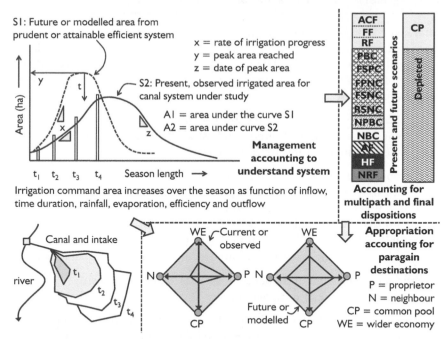

SI: Future or modelled area from prudent or attainable efficient system

x = rate of irrigation progress
y = peak area reached
z = date of peak area

S2: Present, observed irrigated area for canal system under study

AI = area under the curve SI
A2 = area under curve S2

Area (ha)

Season length →

t₁ t₂ t₃ t₄

Irrigation command area increases over the season as function of inflow, time duration, rainfall, evaporation, efficiency and outflow

Management accounting to understand system

ACF
FF
RF
PBC
FSPC
FPNC
FSNC
RSNC
NPBC
NBC
AF
HF
NRF

CP

Depleted

Present and future scenarios

Accounting for multipath and final dispositions

Canal and intake

river

t₁
t₂
t₃
t₄

WE — Current or observed

N ◄ — ► P

CP

WE

N ◄ — ► P

Future or modelled CP

Appropriation accounting for paragain destinations

P = proprietor
N = neighbour
CP = common pool
WE = wider economy

Figure 50 Modelling and measuring system efficiency

Table 24 Three types of water accounts and equivalence in financial accounting

Water accounting type	Equivalence in financial accounting	Notes and application
Water accounting for 'fractions' (or disposition accounts)	Statutory accounts or final accounts	These are relatively simple accounts to show a profit and loss position at a final position e.g. after 12 months. They can be used at irrigation and basin scales, but at the moment are commonly applied to the river basin. Figure 50 includes multipath accounts and the IWMI-WA final outcome. It could also include the Perry framework (2007).
Management accounting	Management accounts/ing	These assist in the ongoing management of a scheme, allowing managers to understand how the 'final' or 'year-end' accounts were achieved. These can be applied to both irrigation system and basin, but with an emphasis on the lower irrigation level.
Appropriation accounting (tracking who gets the paragains)	Profit and loss appropriation accounts/ing	In finance this term is employed to record how profit is dispersed to different 'debtors' such as a reserve account, shareholders, tax payment and reinvestment. This can be done at year-end or on regular basis. Appropriation accounting could be applied to the irrigation system and basin scale.

offer some reflections on one method I have started to develop. It recognises that a practical gauge of system efficiency is the rate of irrigation area completion compared against a modelled ideal rate of area completion (Lankford and Gowing, 1996b). The computation is derived from the continuity equation that relates time to losses:

(applied dose of irrigation, mm depth) = (flow rate, l/sec) ×
   (time taken to irrigate, hours) × (classical efficiency as 'x' per cent) ×
   (0.36) / (completed area irrigated, hectares)

By controlling for, or measuring, irrigation dose, flow rate and flow duration, then one can see that efficiency is proportional to the rate of area completed or volume of water withdrawn or consumed. The attainable efficiency of irrigation (Lankford, 2006) can then be calculated by measuring or modelling what should be possible with reasonable minimum losses (termed prudent irrigation) with what is observed in the system being studied, as the following equation shows:

Efficiency = prudent or modelled volume of water withdrawn/
   observed volume of water withdrawn

By referring to Figure 50, in an irrigated area as with a fixed water supply rate, higher losses express themselves via: a slower rate of progress of irrigation when wetting up at the start of the season (x); a lower total peak area irrigated (y); a later date for this peak (t); a slower rate of ceasing irrigation (z). More efficient irrigation gives a higher rate of irrigation completion, a higher total area of irrigation, an earlier peak and a more rapid cessation of irrigation. At the same time, measures of evaporation, rainfall, average inflow to the irrigation system and any tail-end outflows are required. The variables needed to build the 'prudent' model come from observation, measurement or by interview of farmers. Prudent measures can come from farmers in tail-end part of the irrigation system who manage with very little water. Furthermore calculating the efficiency using the area method allows efficiency to be triangulated using other computations of area, flow, time and depth variables.

The method described refers to catenating irrigation typical of rice systems that spread out from a starting point at the start of the season. These are typical of savannah systems seen in southern Tanzania. Area and scheduling rates for rotational systems (e.g. found in fixed command irrigation systems) would use a slightly modified approach of comparing an observed rate of scheduling and leadstream rotation with a prudent modelled rate.

With regards to appropriation accounting, it should be possible to set up a quantitative model that isolates out material gains and locates their destination as compared to a base-line or expected scenario. Figure 50 schematically contrasts a 'current' quadrangle parasystem on the left with a future distribution on the right. Further development would identify whether the same facilities and metrics for determining disposition and management accounts suffice for appropriation accounts.[6]

7.4 Climate change mitigation, adaptation and resilience

The role of efficiency in climate change mitigation and adaptation requires more room than can be given here and in the light of that some brief comments are offered. The paracommons points to a number of issues when assuming that natural resources efficiency can be raised to either reduce carbon/greenhouse emissions (mitigation) or to deal with the outcomes of increased greenhouse emissions (adaptation). Foremost is the argument that efficiency complexity underwrites expectations for making headway in both of these endeavours.

Taking mitigation first, the paracommons – including the Jevons Paradox – more accurately frames attempts to raise the efficiency with which emissions are reduced or offset via carbon sequestration (see also Sinn, 2012). A more efficient conversion would, the rest being equal, sequester more carbon dioxide for a given input of carbon dioxide (as a dimensionless ratio) or for a given input of land, labour or finance (giving a productivity ratio). Although technologies can deliver these conversions, the paracommons employs nested scales and a longer timeframe in order to verifiably ascertain whether the technology in question truly delivers an aggregate reduction in atmospheric carbon dioxide set against sustaining or increasing parallel societal services or goods. It is via a liminal, complexity lens that optimistic efficiency and traceability claims for carbon sequestration by soil biochar (Lehmann, 2007; Leach *et al.*, 2012) can be judged. What works in the computer, laboratory, experimental station, pilot test, or privately controlled forest plantation, may not apply to 'messy', people-featured, financially-constrained complex and layered savannah and rainforest ecosystems.

In terms of climate change adaptation (CCA), the paracommons suggests that dealing with the symptoms of a changing climate via efficiency may be problematic, and a different order of complexity than responding to CCA via productivity (Mainuddin *et al.*, 2012) concentrating on boosting yields rather than 'saving' water. If we take water as a sector influenced by climate (via for example the increased coverage, duration and severity of drought) then it is difficult to deny that a more efficient use of water makes sense. In such an environment, greater efficiency is likely to underpin the boosting or maintenance of agro-industrial production while consuming less of a drought-scarce resource. While this view is perhaps 'undeniable' for a given single system in a given location at a given time, the means to deliver this outcome, nested within and alongside other systems that are to be 'all efficient' (otherwise the aggregate impact from one part of the system is minimal) is much more problematic. The paracommons reminds us that sector-wise efficiency gains are much more risky than gains delivered from one technology or one unit. Furthermore, gains are likely to fall unequally on competing users. Without substantial care, efficiency-minded adaptation may simply drive up aggregate consumption and further accentuate imbalances of water supply and demand between multiple users.

The topics of climate change resilience and adaptation can be addressed through the lens of paragain distribution, asking who gets the gains arising from a drought-stimulated efficiency change. In the discussion above, using the quadrangle from Figure 22, four main destinations were identified that can receive these gains. Yet *within* those destinations, efficiency gains play out in three different ways that might be seen as flexible/resilience adaptations. These three ways are discussed using one example of a proprietor system; a hypothetical large scale canal irrigation scheme. (Before these are described, recall the prior 'inefficient' system has an extrinsic fraction that either non-beneficially wastes to the atmosphere or sinks, or is recycled to downstream parties.) The three 'resilience' ways that a paragain can be employed within the proprietor system are:

- Storage: the irrigation system can place its paragain within a storage body allowing it to meet a future shortfall in supply.
- Varying the command area: this tactic uses the paragain to extend the irrigated command area during a wet period which then has to shrink when the drought arrives. In both periods, however, the command area is larger than it would be if the system had not raised its efficiency (assuming the paragain is utilised in this manner)..
- Varying the depth applied per irrigation: this tactic changes the depth per irrigation dose which introduces deficit irrigation during the period of shortfall. In parts of Asia this is also known as protective irrigation, spreading a limited amount of water to as many farmers as possible predicated on a reduced volume per farmer. This is likely to ensure that the proprietor and its neighbours all share the drought-induced paragain equally.

The first one, of banking an efficiency gain into storage, is perhaps the better known of the three.[7] The final two could also be characterised as 'non-equilibrium' options that contrast with the first 'equilibrium' approach that aims to flatten out production using a storage buffer. As Leach *et al.* (2010) and Lankford and Beale (2007) argue, purposively allowing for and facilitating variation between dryness and wetness is interpretable as 'resilience', as this quote from the latter's conclusion (p.179) infers:

> Overall, we conclude that sustainable management of variable water resources needs to be grounded in an understanding of the impacts across a river basin; be based on decreasing sensitivity and increasing resilience to water shortage in livelihoods and ecosystems; and continue to see and support water management at the local scale from the viewpoint of closely involved users, while at the same time establishing frameworks at the basin scale that recognise water variability, multiple demands and the manageability of transitions between states of water supply. Facilitating transitions closely adheres to the notion of transformability (Walker *et al.*, 2004) as a determinant of resilience in sustainability.

The three tactics have been exemplified by their application to the proprietor irrigation system. These may also be applied to the other four apexes of the distribution quadrangle in Figure 22. For example, storage could also be used to release water for environmental flows, supply downstream hydropower (the wider economy) or meet the needs of immediate neighbours impacted by a reduction in recycled flows. Taking the whole quadrangle, the parasystem, again the three choices of storage, areal variation and dosage variation apply. In the storage case, water savings from the proprietor system can be banked in a dam (or retained in an aquifer) that is then used for all sectors within the catchment. Furthermore, the paragain need not only stem from one proprietor system, but be summoned by increases to efficiency across all the sectors involved. In terms of the non-equilibrium sharing of a paragain through command area or dosage fluctuations, while this is much more complex, Lankford and Mwaruvanda (2007) show how varying flows within a catchment could be designed into infrastructure to assist in this endeavour.

7.5 Chapter review

Policy to deliver higher efficiencies of complex systems or whole sectors are characterised as being typically either naïve and under-designed or purposively appropriative of resources by powerful players, including the state. Furthermore, the incorporation of efficiency and paracommons thinking into other policy areas of environmental change and sustainability is far from complete. This chapter has alighted upon a few of those aspects such as resilience and ecosystem services, drawing the conclusion that efficiency adds complexity. It does this because efficiency brings additional cyclicality, interconnectedness and recursivity into resource systems and their behaviours. Rising demand in a world alongside an increasingly dynamic, limited supply of natural resources creates new emphases on the role of losses in squaring that circle – yet the theorisation of how efficiency weaves into resilience, adaptation, allocation and other responses is far from complete.

Chapter 8

Conclusions

8.1 A brief review

This book has explored the technological, physical and managerial dimensions of losses, wastes, wastages and efficiency associated with the appropriation of natural resources for human consumption. In considering these dimensions, the manner in which resource efficiency complexity (REC) renders the 'paracommons' was discussed. The paracommons are liminal, neighbourly and redistributive; they contain *potential and multiple* pathways and paradoxes lying between the expectations for, and the outcomes of, intentions to reduce, avoid, recapture, compensate and redirect 'losses' that then provide salvaged resources to be competed over. Competitors comprise users and uses within four parties identified as the proprietor, immediate neighbour(s), the common pool and the wider economy.

The paracommons contain 'spaces of potentiality' prior to the environmental and productive outcomes that resources decompose into once used, consumed, reduced, recovered or avoided. Within the paracommons are located forwards and backwards connections between the pre- and post-conversion of resources, including efficiency technologies to be enacted, institutions and differing knowledges. These connections influence the location and availability of the principal resources and post-liminal resources, goods and paragains. The liminal paracommons offers a pause prior to forming overly optimistic prefigurations of new productivities and savings of common pool resources.

Resource efficiency complexity, it is argued, arises from 20 sources, which in turn express themselves into three types of liminal spaces: 'productive-consumptive', 'multipath paracommons' and 'multiservice paracommons'. The productive-consumptive liminal space is best exemplified by hydrocarbon fuel efficiency. The multipath space describes water and irrigation within river catchments. The multiservice space is exemplified by the transformations of carbon and carbon dioxide fluxing between atmosphere, forests, other biomass and soil, and their impacts on other ecosystem services.

The significance of the paracommons and resource efficiency complexity is correlated to the magnitude of the scarcity, value and complexity of the loss fraction associated with the principal resource. Thus with irrigation the paracommons is 'made real' because the loss fraction is significant, valuable and fluid. Poorly theorised and executed reworking of complex systems result in high levels of uncertainty surrounding the kinetics and destinations of the losses, wastes and wastages. The liminal spaces associated with the paracommons insert themselves between the state and irrigator in the form of takings of wastage fractions; between the disciplines and languages of politics, law, economics and engineering in how they frame consumption and efficiency, and between judgments of 'best, good, optimal, poor' resource management. The paracommons is at the heart of political, technological and regulatory reforms designed to conserve resources or boost productivity and as such, would surely be somewhere in a future articulation of Linton's 'socioecological nature of water' (2008, p.646) and 'hydrosocial cycle' (Budds, 2009).

8.2 Inserting efficiency into the governing of the commons

The paracommons and REC throws light on a critical revisiting of the modernisation of nature started by Hays in 1959 and continued by Clark and York (2005) and others that one of the chief aims of environmental governance is to protect and sustain nature while meeting the economic needs for the growing human population of the planet. This implies a building up of the comprehensiveness with which we govern the commons, drawing in resource use efficiency and its implied technologies and institutions.

There is much that can be said here on the politically problematic mechanisms for achieving sustainable environmental governance such as green capitalism. By problematising efficiency and productivity within this modernisation trajectory, the paracommons points to profound and multiple sources of 'systems uncertainty'. In contrast to the (in my view) conventional framing of nature-serving-society as through an ecosystem services prism, it is possible to see that resource and waste/wastage flows create a complicated nested and recursive embedding of social-ecological-technological-systems. In these systems, resources and paragains flow, cascade, interact and switch, are attractive to some and neutral or harmful to others, and are in quantities and qualities that change over time and space. These 'agro-eco-built-systems' vary in type, from the engineered 'end' of resource use (distributing water via canal systems, arguably containing freshwater away from 'nature') to forestry, watersheds and carbon sequestration that shift questions of productivity and efficiencies to 'within nature' (albeit where technologies of forestry husbandry and harvesting exist).

Ecologists might argue that nature cannot be inefficient given that all energy, light, nutrient and enzymatic flows cascade through ecology to produce myriad species, interactions and services (e.g. Phillips, 2008). Yet efforts at REDD+ and payments for ecosystem services (García-Amado et al., 2011) signal that our credo is to restore degraded or sub-performing environments by managing the performance and products of natural conversions. Yet even this productivity-boosting industrialising emphasis of eco-efficiency insufficiently captures the efficiency complexities at the heart of ecosystem services. The emphasising of some services alongside a de-emphasising of others switches the denominator – resource users previously compelled to sustain one service find themselves in a liminal space as their 'service' is now deemed less important and more of a 'loss'.

Yet there remains the normative and entirely worthy project of seeking to do better with socio-ecological systems to make them (relatively) more efficient and productive in the future than they are today. This book would not be written if its author did not believe that irrigation systems and sectors could be made more productive and less consumptive via engaging with efficiency. However, caution immediately applies as to how to achieve that. The route to thinking about this is surely a social, deliberative one: to address the multiple views on performance arising amongst users and disciplines, identified long ago by Murray-Rust and Snellen (1993). Devising methodologies for this deliberative endeavour that work effectively may be equally as complex (Scoones, 1999).

Summarising, the paracommons stems from differences between policy prefigurations and physical outcomes arising from the complexity of 'flows' of resources and their loss fractions where the latter are deemed valuable for a variety of users and sectors. In responding to this policy uncertainty, resource efficiency complexity and the paracommons corrals and suggests seven concerns about resource governance and management:

- First, is the observation that regulatory instruments (legal, market and customary) for the ownership and regulation of recycling and losses are being outstripped by fast-moving events driven by scarcity, necessity and technical ingenuity. If not regularly updated and reformed, normative procedures for managing resources also sitting within disciplines such as engineering and law, will fail to advance productivity gains while attending to environmental, social and sustainability criteria.

- Second is the expectation that the physical parameters of natural resources need to be better quantified and understood. The tracking of water in river basins and irrigation is an omission in CPR and co-management research. Another example of measurement omission is in the emerging claims of water offsetting, neutrality and virtual water transfers (Hoekstra, 2008; and criticisms of this work by Wichelns, 2011). It is my belief that much of this work is dependent on modelled assumptions. Yet even with better monitoring, cautions nevertheless apply. The paracommons framing suggests that research of causality between efficiency interventions and outcomes may not be possible unless a more comprehensive picture of resource flows is gathered (see also Sorrell and Dimitropoulos, 2008).

- Third, because resources such as water, carbon and energy inter-connect, their wastage fractions require heightened scrutiny so that externalities and impacts of the consumption of one resource on another are educed.
- Fourth, and connected to the first three concerns, is the expectation that the frames of resource governance and common pool theory will have to widen if we are to understand the liminal and pre- and post-liminal products and processes (including those reduced and avoided) of primary resource consumption. The evidence from irrigation is that the paracommons not only determines access to the salvaged resources, but that these feed 'backwards' into mechanisms that determine the scheduling and distribution (and therefore performance) of the original principal resource. In this way, with increasing scarcity, 'losses, waste and wastage' ordinarily seen as at the margins of resource systems move to the centre ground, shaping debates via the accounting methods adopted. It is for this reason that I believe the paracommons is best addressed via a more comprehensive science of resource efficiency complexity and kinetics.
- Fifht, as Knox et al. (2012) show (also supported by Boelens and Vos, 2012, and Lankford, 2012b) the ways and means of including resource users are likely to comprise a rewarding part of future efficiency research and, out of the results from that work, a rich new source of understandings. Including users at the centre of the efficiency may have profound consequences for the choice of policy and modelling of irrigation efficiency improvements. This will shape new priorities in water science – defining and evaluating efficiency measures that support resource users in terms they understand.
- Sixth, this analysis asks what comprises 'the commons' or indeed 'ecosystem services'. In this debate, this book suggests that in a closing-system world where wastes/wastages are vested with virtue and natural resources and wastes/wastages invert and swap places, the commons and ecosystem services are redefined by new levels of scarcity, efficiency, recovery and avoidance. The carbon dioxide 'commons' converted to overlapping ecosystem services is one example of this phenomenon, and irrigation losses as the remaining source of water in a closed river basin is another. Whether categorisation of ecosystem services (supporting, provisioning, etc.) will stand further efficiency-related scrutiny is also a question for future work.
- Seventh, the terms and ideas employed for efficiency and productivity will necessarily continue to evolve in order to capture useful expressions of ratios between inputs and outputs. We can see this in the irrigation literature. There is a tension here, however. First one must ask whether this evolution is taking place fast enough and is being communicated to resource stakeholders comprehensively enough. In the irrigation sector, many stakeholders remain blissfully unaware of the 15-year debate on efficiency (Molden et al., 2010). Second, evolution of terms might usefully recognise the linguistics of waste and efficiency. Attempts to rationalise a single taxonomy of resource use (Perry, 2011) may be too premature.

8.3 Concluding thoughts

Although this book explores resource efficiency in depth, I am the first to see some gaping holes – prime of which is the lack of field data to support a fuller working of the paracommons concept. I also suspect this attempt at exploring the efficiency of use of natural resources will not meet everyone's favour. I say this because it is the very nature of efficiency and its surprising complexity that allows our disciplinary training and experience to view it in different ways. The identified factors of resource efficiency complexity are without doubt incomplete or inappropriately arranged depending on the reader's training and interests. As van Halsema and Vincent (2012) allude, it is the multi-spectrum mosaic of scales, technologies, institutions and interventions that may feed disagreements over theory and praxis and in doing so provide rich pickings for political ecologists analysing discourses on resource use efficiency. In predicting a future political ecology of resource efficiency and productivity, I characterise this will not be working on 'new frontiers' but instead on consolidation, informed by empirical findings, between strands of existing formulations and actors involved, as rich as that may be.

Furthermore, I am dissatisfied regarding the depth and detail of some of my discussion. I have played to my 'irrigation' strengths while cueing up an introduction to efficiency complexity. As such, I have not specified a comprehensive, cross-disciplinary approach to managing the paracommons and the politics and technologies of efficiency within resources conservation and ecosystem services. And this book will leave the discussion on REC implications for climate change adaptation and resilience to others.

Nevertheless, I strongly agree with Neuman's (1998) recommendation for a more systematic approach to improving efficiency and performance, and placing this challenge at the heart of sustainable productive and equitable/just resource management. Given that issues of scarcity, recycling, aggregate depletion and linkages between resources are increasing, the phenomenon of a commons of losses, waste and wastage seems set to rise in prevalence. The Montana vs Wyoming dispute over the ownership of previously-wasted now-conserved water (Norris, 2011) may be a foretaste of future competition over material gains 'freed up' from efficiency gains. The concomitant response is to create efficiency-cognisant epistemic communities of scientists, users and service providers capable of addressing efficiency-induced complexity, thereby shrinking the degree or opaqueness of liminal uncertainty. Knowledge and information will be key because it is my view that the delivery of more efficient, less consumptive parasystems will be localised and specific to each case due to the individual complexities found therein.

Or being even more critical, those studying access, equity, sufficiency, justice and productivity of resource distributions might explain why the efficiency-centred ideas on these pages either need rejecting or reshaping before their significance for resource governance can be explicated. Perhaps future work will take several of the concepts posited here and re-arrange them to reveal new ways of looking at efficiency and its relationship with economic efficiency and other measures of resource distribution and redistribution (Storper, 2011).

That said, this analysis reflects the insufficient and incomplete research applied to irrigation as a whole (and by extension to ecosystem services). It is a decalculated field or to be more accurate, it is an over-modelled, narrative-driven yet empirically impoverished field. It is my view that irrigation systems science would benefit from much more field research. Furthermore, as well as mapping out the multiple dimensions of efficiency and productivity, work is required to identify the practical innovations for boosting them (while reconciling system and basin scales). Although attributable, accurate connections between theory, measurement, practice and context may not be readily forthcoming, effective policy that delivers productive, equitable and less consumptive irrigation should be the yardstick by which we measure theoretical and methodological progress.

I surmise that with an unfolding set of concerns around food, water, energy, climate change and freshwater biodiversity, the topic of consumptive and non-consumptive use of resources, and the distribution and destination of constituents such as 'waste, wastage and losses', mediated by engineers, scientists and resource users, will be of profound significance in the next few decades. The functional calculation of the large volumes of water transferred through irrigation hints at rewards to come but runs the risk of masking the multitudinous questions associated with those calculations and their formulation. Similar questions and doubts arise in the field of the ecosystem services and efficiency of carbon emissions, offsetting and sequestration. In this progression, we might witness a number of shifts: a rising tempo of scientific interest in resource efficiency, a migration of the discussion of efficiency from academia to the public and policy domain, and a new application of common pool science and studies to the material gains surrounding the efficiency of resource utilisation. Enhanced control over efficiency transformations moderates beliefs that higher efficiencies lessen resource consumption while addressing the paradox of efficiency-induced increased consumption. New ways of thinking about efficiency would benefit sustainable food production, economic growth and a more effective protection of ecosystem services.

Notes

1 A preliminary explanation of the paracommons

1 'The Federal Water Minister admits new tax legislation has to be finalised so that Murrumbidgee Irrigation can take up funding for water-saving infrastructure projects. On Friday the government announced the Murray Darling Basin Plan will deliver an extra 450 billion litres of water to the river system, mainly through on-farm irrigation efficiency projects'. http://www.abc.net.au/news/2012-10-29/basin-burke/4338574?&section=news. Accessed 29 October 2012. See also Australia's 2012 Factsheet on water saving infrastructure. http://www.environment.gov.au/water/publications/mdb/pubs/fact-sheet-water-saving-infrastructure.pdf. Accessed 30 October 2012.

2 The Supreme Court decided that the under the 'prior appropriation' legislation operating in Western USA and applicable to both states, Montana irrigators were allowed to change irrigation practices despite these now resulting in smaller recoverable losses now no longer flowing to Wyoming. In other words, the proprietor got the gain. See Chapter 6 (Behnampour, 2011; Bell, 2007; Neuman, 1998).

3 Las Vegas Review Journal. 'Colorado River water pact could be model for other nations'. Posted: Nov. 15, 2012. Accessed 17 November 2012. http://www.lvrj.com/news/colorado-river-water-pact-could-be-model-for-other-nations-179573071.html.

4 Also http://www.circleofblue.org/waternews/2012/world/u-s-mexico-sign-major-deal-on-colorado-river-issues-delta-restoration-infrastructure-water-sharing/ Accessed 10 December 2012.

5 I take the commons and the paracommons as singular; 'it' rather than 'they'. I specify types of paracommons and use the word generically, as happens with 'the commons'; authors write about individual commons (e.g. fisheries) as well as 'the commons'.

6 Here the idea of 'parallax' supports the use of the prefix 'para'. While writing this article, I read in the *Guardian* (UK newspaper) on Tuesday 3 July 2012 an article showing this kind of relationship: that the establishment of a waste incinerator plant would deprive waste dump pickers of their livelihoods; http://www.guardian.co.uk/world/2012/jul/02/future-for-india-waste-pickers

7 Examples of holistic treatments include ecosystem services, integrated water resources management, (IWRM), the livelihoods framework, the study of socio-ecological systems (SES), and the study of large technical systems (LTS).

2   Main introduction and the scope of the book

1   Tare originates from the Arab word *tarha*, meaning 'what is rejected' (OED, 2012).

2   Following dictionary definitions, 'wastes' are usually undesirable by-products and wastages are unrecoverable losses during conversion. For irrigation, saline drainage water is waste and soil evaporation is wastage. This distinction rarely holds: waste takes place because of wastage, and depending on what is valuable in the future, they interchange. Furthermore 'wastes' often have a negative connotation – hence the difference between wastewater irrigation and inefficient irrigation. These complications help define the uncertainties of the paracommons.

3   A number of references to the lack of flow measurement on irrigation schemes are made in this book. I believe the vast majority of the world's gravity-fed irrigation schemes are without satisfactory levels of monitoring and measurement of flow rates and other variables. A debate on this goes back many years, see, for example, Kraatz and Mahajan, 1975; Replogle, 1989; Lankford, 1992).

4   Unless clarified, the term efficiency includes the idea of raising resource productivity.

5   Nevertheless, the imprint of a dynamic natural world exacerbates the irresolution of complex efficiency conversion processes at the heart of the paracommons concept.

6   I understand the paracommons as an abstract form of the commons. However, the distinction might not hold under more scrutiny in the future; indeed the commons is also arguably an abstract form of physical entities such as forests, rivers and oceans.

7   At the moment, declining efficiency is not explicitly addressed by this book.

8   I've drawn on some systems literature (Wilson, 1990; Checkland, 1999; Skyttner, 2006) but mindful of space and word limits, I have not addressed this directly.

9   I wrote this down in 2008 as a transcript of a British TV documentary but failed to find further reference to this. President Carter appeared to be at an official hearing.

10   For example, in 2011, I was sent the results of a project by a global environmental organisation that explored efficiency in Indian irrigation systems. I could not make sense of their data. I also have not seen any peer reviews of the work conducted by global drinks manufacturers who claim to have improved the efficiency of their production systems beyond their plants and factories. Furthermore, I have not found any detailed hydrological pre- and post-analysis of the Montana vs. Wyoming case (while I don't deny a change in downstream flows, the legal commentary does not give a full picture). In my attempt to fill this gap, I submitted a large research project to UK research funders in May 2011 to study water productivity and efficiency on irrigation systems in Ethiopia and Tanzania with IWMI and national universities. The project planned multiple measurements to track resources and isolate intermediate and aggregate system efficiencies. It was not funded.

11   A calculation shows that, say, 20,000 hectares of irrigated fields with a ratio of 10 hectares supplied by an average of 1000 metres of open canal of about 0.75 metres width will evapotranspire water in a similar ratio of canal area surface area to irrigated field area; 150 to 20,000 hectares. Put another way, 0.75 per cent of 20,000 hectares of evapotranspiration comes from canals. And yet, the Australian Government 'Factsheet on Water-saving infrastructure in the Murray-Darling Basin' (downloaded 2012) informs readers, alongside a photo of large bore pipes, that 'Replacing irrigation channels with pipelines can reduce water loss through evaporation'. Have the authors confused water control with water losses?

3   On resource efficiency; multiple views

1   This high rate of evapotranspiration, by open crop stomata, can be nearly equal to open water evaporation. It means that irrigation, particularly when extending to thousands of hectares per river basin, is highly consumptive of water.
2   There is an extensive literature on design of water control: Plusquellec *et al.* (1994); Renault *et al.* (2007).
3   My understanding is these models do not see themselves as schools, and future classifications might depict different types or numbers of schools.
4   Although recoverable drainage flows quickly in ditches, it is the passage of this fraction through an irrigation system and its soil that takes time.
5   When working on sugarcane irrigation systems in the 1980s, I recorded flood irrigation efficiency and uniformity to be higher than sprinkler efficiency particularly during hot dry windy days. The scheme was 'blessed' with cracking heavy soils that reduced deep seepage after sealing allowing surface irrigation to be surprisingly efficient. In one test, I recorded a classical efficiency of 90 per cent up to the secondary canal level. It was so different from received wisdom about surface irrigation that I thought I had made an error.
6   Here the term 'efficiency' is of its era, aligned to interests in throughput and 'modernisation'. Thus although I engage with environmental productivity, I am not promoting environmental maximisation – see Schroeder (2000).
7   Search online for World Water Week, 28 August 2012 Workshop K12: Best Use of Blue Water Resources for Food Security.
8   Beyond email exchanges between individuals, it is not easy to obtain definitive evidence of IWMI, ICID and other organisations deciding not to use 'efficiency' but this quote in the Perry 2007 abstract summarises an ICID viewpoint 'ICID recommends that this terminology be used in the analysis of water resources management at all scales, and form the basis for its research papers and other published outputs.' Then on page 372, the author hints at the tensions in the deployment of the word 'efficiency': 'Given the contribution that IWMI made, both in supporting the proposals of others to use less confusing terms and in its own proposals, as summarised above, it is surprising that the "efficiency" word should reappear'.
9   Text from Lankford (2012, p.34) exemplifies:

> There exist confusions over modes of modernity in irrigation, with different types being strongly associated or even representing levels of waste. 'Traditional' implies considerable waste while 'modern' relates to reduction in waste: 'Even in the mainly urbanised Damascus basin, however, around 80 percent of available water is used in agriculture, with outdated irrigation methods wasting huge quantities of it. Across Syria as a whole, only 16 percent of farmers use modern irrigation systems, according to JICA's Mori.'
>
> (IRIN, 2010)

> While this quote does not specify a definition of 'modern', further investigation makes clear that Syria defines modern technology as drip and sprinkler systems (Varela-Ortega and Sagardoy, 2003).

10   Given the scope of the book I stick with examples of efficiency and productivity that are relatively simple – 'x' per cubic metre or 'y' per hectare. Qualitative scores and more complex quantitative ratios of ecosystem services are not explored – for example, numbers of biological species sustained per timber tonnage harvested, or streamflow runoff (m^3) per unit of carbon (kg) sequestered.

11 Water use efficiency is now being increasingly adopted for sector wide thinking rather than the narrower crop application – for more on this see van Halsema and Vincent (2012).

12 Al-Mashaqbeh *et al.* (2012) use the word 'conserve' in the same way: 'grey water systems as an opportunity to conserve potable quality water' (p. 581).

13 In the paper by Spronk (2010) the unwary reader might believe efficiency relates to wastes and recycling rather than efficiency in the economic or social sense.

14 See http://www.waterwise.org.uk/

15 At the time of writing, plans were being developed by IFC to address irrigation efficiency via a 'Global Efficient Irrigation Program' (GEIP). Partnerships with private micro-drip manufacturers comprise such initiatives.

16 Curiously, there appears to be greater research interest in wastewater irrigation (using poor quality urban wastes) than inefficient irrigation (producing or consuming re-usable losses); see Scott and Raschid-Sally (2012).

4   A framework of resource efficiency complexity

1 One interpretation: I do not deny that alternative arrangements and factors exist.

2 Recovered usually means how waste (by definition easily recovered) is recycled: for example, household plastics are recovered. However, it could also mean the 'reducing' of wastage (by definition not easily recovered, for example, household heat-loss) by a new recovery technology, achieving a 'recovery' previously not possible. If, however, this heat loss was reduced by better insulation, this would be 'forestalling'.

3 Table 10 primarily addresses the topic of losses through the lens of water management. Regrettably there is insufficient space to develop a taxonomy of intrinsic and extrinsic fractions in ecosystem services.

4 For example, new technologies for heat recovery from low temperature steam.

5 I envisage arrangements where the 'client' of a proprietor is also its neighbour – for example, in industrial ecology, one factory sends waste to another in a sequence.

6 Regrettably space does not allow exploration of the fascinating topic of rewarding efficiency (see 'negawatts', or the incentives to save, for example, explored by British Colombia BC Hydro – although the latter offers mainly technology subsidies).

7 It is this expansion that shows how a more efficient proprietor would send water to a neighbour if the neighbour occupied the 'to be expanded' land.

8 For the time being I choose not to explore efficacy and effectiveness.

9 Daly and Farley (2011, pp.165–167) comment on the effectiveness of many people working on a problem via 'the Linux operating system and open source software: the efficiency of the information commons.'

10 In an unpublished 2012 systematic review of the water allocation literature for the UK's Department for International Development, in which the author was involved, a search of approximately 30,000 publications revealed that about 30 contained accurately derived and analysed data on the physical outcomes of policy.

11 I include the United Nations System for Environmental-Economic Accounting for Water (UNSEEAW) (UNSD, 2007) with the Perry approach.

12 As understood by the author arising from various conversations during 2011 with Chris Perry, he does not see these fractions as pathways.

5 The liminal paracommons

1 Following Perry (2007), changes in water storage (e.g. soil moisture content) would make up a seventeenth pathway but is excluded from this framework.
2 Sikor (2013) points out that the United Nations Framework Convention on Climate Change (UNFCCC) defines safeguards (Decision 1/CP.16) to protect people from adverse impacts of REDD+ projects recognising they forego some ES when turning ecologies towards provisioning services such as palm oil and timber.
3 While studying for a BSc in Soil Science, it was made clear to me that while soil carbon readily oxidises under new intensive land use practices, rebuilding soil carbon even through biochar was a hard-won, long-term effort. See Robbins (2011) regarding the challenge of achieving net sequestration of soil carbon under conventional agricultural practices, also taking into account the carbon costs of inputs used such as nitrogen fertiliser.
4 'Blue' water found in rivers, pipes and canals, takes up different physical pathways with the wastage being vapour (i.e. as a multipath paracommons), whereas rainwater and green water (soil moisture and transpiration) (Falkenmark and Rockström, 2004) offers different multiservice pathways. Replacing virgin rainforest with oil-palm plantations is an example. The latter uses local rainfall/moisture regimes to produce palm oil while eroding biodiversity.

6 Distinctions between the commons and the paracommons

1 This essentialises the principal commons – see Finley (2009), for example, on a more nuanced examination of fish harvest rates.

7 Significances and applications of the paracommons

1 This is a potential risk of the IFC approach to global irrigation efficiency via large-scale micro-drip investments.
2 Similar points apply to the inclusion of views of people affected by REDD+ programmes (Sikor, 2013).
3 The game was part of the DFID-funded RIPARWIN project (Raising irrigation productivity and releasing water for intersectoral needs) (McCartney et al., 2007).
4 At the workshop where Figure 45's photos were taken, local irrigators claimed a large-scale dam in the high catchment would solve their perceived scarcity of water.
5 The story of the wet and dry water saving technology 'found' in rice irrigation in Madagascar might be interpreted along waterist lines (Rappocciolo, 2012).
6 The problem of verifying the relative losses and gains of carbon and different ecosystem services appears to be similar in some respects (Sikor, 2013).
7 Predicting the crop production response to these three options depends upon the biomass-water production curve for the crop and a number of other agronomic factors. To answer this production question and the returns to water, land, labour and energy requires in-depth research.

References

2030 Water Resources Group (2009) *Charting Our Water Future Economic Frameworks to Inform Decision-making*. The Barilla Group, The Coca-Cola Company, The International Finance Corporation, McKinsey and Company, Nestlé S.A., New Holland Agriculture, SABMiller PLC, Standard Chartered Bank, and Syngenta AG. Available: http://www.2030wrg.org/

Acheson, J.M. and Brewer, J.F. (2003) 'Changes in the territorial system of the Maine lobster industry', in N. Dolšak and E. Ostrom (eds) *The Commons in the New Millennium: Challenges and Adaptation*, MIT Press, Cambridge, MA.

Adams, W.M. (2008) *Green Development: Environment and Sustainability in a Developing World*, third edition, Routledge, London.

Agrawal, A. (2005) *Environmentality: Technologies of Government and the Making of Subjects*, Duke University Press Books, Durham, NC.

Al Sabbagh, M.K., Velis, C.A. Wilson, D.C., and Cheeseman, C.R. (2012) 'Resource management performance in Bahrain: a systematic analysis of municipal waste management, secondary material flows and organizational aspects', *Waste Management and Research*, vol. 30, no. 8, pp.813–824.

Al-Mashaqbeh, O.M, Ghrair, A.M. and Megdal, S.B. (2012) 'Grey water reuse for agricultural purposes in the Jordan Valley: household survey results in Deir Alla', *Water*, vol. 4, pp.580–596.

Anderberg, S. (1998) 'Industrial metabolism and linkages between economics, ethics, and the environment', *Ecological Economics*, vol. 24, nos 2–3, pp.311–320.

Antle, J., Capalbo, S., Mooney, S., Elliott, E. and Paustian, K. (2003) 'Spatial heterogeneity, contract design, and the efficiency of carbon sequestration policies for agriculture', *Journal of Environmental Economics and Management*, vol. 46, no. 2, pp.231–250.

ASARECA (2006) Maputo Workshop Statement. Association of Strengthening Agricultural Research in Eastern and Central Africa (ASARECA), available: www.asareca.org/swmnet

Ayres, R.U. and Ayres, L.W. (1996) *Industrial Ecology: Towards Closing the Materials Cycle*, Edward Elgar, Cheltenham.

Bachman, K.L. (1952) 'Changes in scale in commercial farming and their implications', *American Journal Agricultural Economics*, vol. 34, no. 2, pp.157–172.

Baden, J.A. and Noonan, D.S. (eds) (1998) *Managing the Commons*, second edition, Indiana University Press, Bloomington, IN.

Bailey, I., Gouldson, A. and Newell, P. (2011) 'Ecological modernisation and the governance of carbon: a critical analysis', *Antipode*, vol. 43, no. 3, pp.682–703.

Banerjee, S.B. (2003) 'Who sustains whose development? Sustainable development and the reinvention of nature', *Organization Studies*, vol. 24, no. 1, pp.143–180.

Barlow, M. (2009) *Blue Covenant: The Global Water Crisis and the Coming Battle for the Right to Water*, New Press, New York.

Behnampour, L.C. (2011) 'Reforming a Western institution: how expanding the productivity of water rights could lessen our water woes', *Environmental Law*, vol. 41, no. 1, pp.201–230.

Bell, C. (2007) 'Promoting conservation by law: water conservation and western state initiatives', *University of Denver Water Law Review*, vol. 10, p.313.

Berge, E. and Van Laerhoven, F. (2011) 'Governing the commons for two decades: a complex story'. *International Journal of the Commons*, vol. 5, no. 2, pp.160–187

Bhuiyan, S.I. (1982) 'Irrigation system management research and selected methodological issues', *IRRI Research Papers Series*, no. 81, International Rice Research Institute, Manila, Philippines.

Biswas, M.R. (1990) 'Interaction between design and operation and maintenance of farmer-managed irrigation systems', in R. Yoder and J. Thurston (eds), *Design Issues in Farmer-Managed Irrigation Systems: Proceedings of an International Workshop of the Farmer-Managed Systems Network*, organised by the International Irrigation Management Institute and The Thailand Research on Irrigation Management Network and held at Chaing Mai, Thailand from 12–15 December 1989, International Irrigation Management Institute, Colombo, Sri Lanka, pp.193–205.

Blaikie, P. and Brookfield, H. (1987) *Land Degradation and Society*, Methuen, London.

Bleischwitz, R., Welfens, P.J.J. and Zhang, Z. (eds) (2011) *International Economics of Resource Efficiency: Eco-Innovation Policies for a Green Economy*, Springer, Heidelberg.

Boelens, R. (2008) 'Water rights arenas in the Andes: Upscaling the defence networks to localize water control', *Water Alternatives*, vol. 1, no. 1, pp.48–65.

Boelens, R. and Vos, J. (2012) 'The danger of naturalizing water policy concepts: water productivity and efficiency discourses from field irrigation to virtual water trade', *Agricultural Water Management*, vol. 108, pp.16–26.

Bolding, A., Mollinga, P.P. and Van Straaten, K. (1995) 'Modules for modernisation: colonial irrigation in India and the technological dimension of agrarian change', *The Journal of Development Studies*, vol. 31, no. 6, pp.805–844.

Boons, F. and Howard-Grenville, J. (eds) (2009) *The Social Embeddedness of Industrial Ecology*, Edward Elgar, Cheltenham, UK, and Northampton, MA.

Borgia, C., García-Bolaños, M. and Mateos, L. (2012) 'Patterns of variability in large-scale irrigation schemes in Mauritania', *Agricultural Water Management*, vol. 112, pp.1–12.

Bos, M.G. (1979) 'Standards for irrigation efficiency of the International Commission on Irrigation and Drainage (ICID)', *Journal of the Irrigation and Drainage Division of the American Society of Civil Engineers (ASCE)*, vol. 105, no. 1, pp.37–43.

Bos, M., Kselik, R.A.L., Allen, R.G. and Molden, D. (2009) *Water Requirements for Irrigation and the Environment*, Springer-Verlag, New York.

Bourg, D. and Erkman, S. (2003) *Perspectives on Industrial Ecology*, Greenleaf Publishing. Sheffield.

Brede, M. and Boschetti, F. (2009) 'Commons and anticommons in a simple renewable resource harvest model', *Ecological Complexity*, vol. 6, no. 1, pp.56–63.

Bretschger, L. (2011) 'Sustainability economics, resource efficiency, and the green new deal', in R. Bleischwitz, P.J.J. Welfens and Z. Zhang (eds) *International Economics of Resource Efficiency: Eco-Innovation Policies for a Green Economy*, Springer, Heidelberg.

Bruns, B. (2011) 'Crafting rules for an invisible commons: responding to Yemen's groundwater grisis', presentation at the University of North Carolina, Chapel Hill, 27 March 2011.

Buckingham, S., Marandet, E., Smith, F., Wainwright, E., and Diosi, M. (2006) 'The liminality of training spaces: places of private/public transitions', *Geoforum*, vol. 37, no. 6, pp.895–905.

Budds, J. (2009) 'Contested H_2O: science, policy and politics in water resources management in Chile', *Geoforum*, vol. 40, no. 3, pp.418–430.

Cadenasso, M.L., Pickett, S.T.A. and Grove, J.M. (2006) 'Dimensions of ecosystem complexity: heterogeneity, connectivity, and history' *Ecological Complexity*, vol. 3, no. 1, pp.1–12.

Campbell, B. (2013) *Living Between Juniper and Palm: Nature, Culture and Power in the Himalayas*, OUP, Delhi.

CAWMA (2007) *Water for food, Water for Life; Challenge Programme on Water and Food. Comprehensive Assessment of Water Management in Agriculture (CAWMA)*, Earthscan, London, and International Water Management Institute, Colombo.

Chambers, R. (1988) *Managing Canal irrigation: A Practical Analysis from South Asia*, IBH Publishing, New Delhi and Oxford.

Chartres, C. and Varma, S. (2010) *Out of Water: From Abundance to Scarcity and How to Solve the World's Water Problems*, FT Press, Upper Saddle River, NJ.

Checkland, P. (1999) *Systems Thinking, Systems Practice*, John Wiley and Sons, Chichester.

Chen, D., Webber, M., Chen, J. and Luo, Z. (2011) 'Energy evaluation perspectives of an irrigation improvement project proposal in China', *Ecological Economics*, vol. 70, pp.2154–2162,

CIAT (2012) *Eco-Efficiency: From Vision to Reality*. Issues in Tropical Agriculture. International Center for Tropical Agriculture (CIAT), Colombia.

Clark, B. and York, R. (2005) 'Carbon metabolism: global capitalism, climate change, and the biospheric rift', *Theory and Society*, vol. 34, no. 4, pp.391–428.

Clarke, A. (2006) 'Seeing clearly: making decisions under conditions of scientific controversy and incomplete and uncertain scientific information', *Natural Resources Journal*, vol. 46, no. 3, pp.571–599.

Clement, F. (2013) 'From water productivity to water security: a paradigm shift?', in B.A. Lankford, K. Bakker, M. Zeitoun and D. Conway (eds) *Water Security: Principles, Perspectives and Practices*, Earthscan Publications, London.

Cole, D.H. (2002) *Pollution and Property: Comparing Ownership Institutions for Environmental Protection*, Cambridge University Press, Cambridge.

Corbera, E., Estrada, M. and Brown, K. (2010) 'Reducing greenhouse gas emissions from deforestation and forest degradation in developing countries: revisiting the assumptions', *Climatic Change*, vol. 100, pp.355–388.

Cox, M., Arnold, G. and Villamayor Tomás, S. (2010) 'A review of design principles for community-based natural resource management', *Ecology and Society*, vol. 15, no. 4, article 38.

Crase, L. and O'Keefe, S. (2009) 'The paradox of national water savings', *Agenda*, vol. 16, no. 1, pp.45–60.

Crase, L., O'Keefe, S. and Dollery, B. (2009) 'Water buy-back in Australia: political, technical and allocative challenges', paper presented at Fifty-third AARES Conference, 11–13 February 2009, Cairns, Australia, Australian Agricultural and Resource Economics Society (AARES).

Cumming, G.S. (2011) *Spatial Resilience in Social-Ecological Systems*, Springer, London.

Daly, H.E. (1992) 'Allocation. distribution and scale: towards an economics that is efficient, just and sustainable', *Ecological Economics*, vol. 6, pp.185–193.

Daly, H.E. and Farley, J. (2011) *Ecological Economics, Second Edition: Principles and Applications*, Island Press, Washington, DC.

Danielsen, K.A., Fulton, W. and Lang, R.E. (1999) 'Retracting suburbia: smart growth and the future of housing', *Housing Policy Debate*, vol. 10, no. 3, pp.513–540.

Davenport, J.H. (2008) 'Less is more: a limited approach to multi-state management of interstate groundwater basins', *University of Denver Water Law Review*, vol. 12, pp.139–180.

Dellapenna, J.W. (2002) 'The law of water allocation in the Southeastern states at the opening of the twenty-first century', *University of Arkansas at Little Rock Law Review*, vol. 25, pp.9–88.

Dietz, T., Ostrom, E. and Stern, P.C. (2003) 'The struggle to govern the commons', *Science*, vol. 302, no. 5652, pp.1907–1912.

Döll, P. and Siebert, S. (2002) 'Global modelling of irrigation water requirements', *Water Resources Research*, vol. 38, no. 4, pp.8.1–8.10

Dolšak, N. and Ostrom, E. (2003a) 'The challenges of the commons', in N. Dolšak and E. Ostrom (eds) *The Commons in the New Millennium: Challenges and Adaptation*, MIT Press, Cambridge, MA.

Dolšak, N. and Ostrom, E. (eds.) (2003b) *The Commons in the New Millennium: Challenges and Adaptation*, MIT Press, Cambridge, MA.

Dolšak, N., Brondizio, E.S., Carlsson, L., Cash, D.W., Gibson, C.C., Hoffman, M., Knox, A., Meinzen-Dick, R. and Ostrom, E. (2003) 'Adaptation to challenges', in N. Dolšak, and E. Ostrom (eds) *The Commons in the New Millennium: Challenges and Adaptation*, MIT Press, Cambridge, MA.

Eichner, T. and Pethig, R. (2011) 'Carbon leakage, the green paradox, and perfect future markets', *International Economic Review*, vol. 52, no. 3, pp.767–805.

Environment Agency (2009) 'Turning the tap on business water waste: Business best practice from the Water Efficiency Awards 2009'. Accessed online: www.environment-agency.gov.uk/water-efficiency-awards.

Falkenmark, M. and Molden, D. (2008) 'Wake up to realities of river basin closure', *International Journal of Water Resources Development*, vol. 24, no. 2, pp.201–215.

Falkenmark, M. and Rockström, J. (2004) *Balancing Water for Humans and Nature: The New Approach in Ecohydrology*, Earthscan, London.

FAO (1977) *Guidelines for Predicting Crop Water Requirements*, FAO Irrigation and Drainage Paper No. 24. Food and Agriculture Organization of the United Nations, Rome.

FAO (1999) *Crop Evapotranspiration*, FAO Irrigation and Drainage Paper No. 56. Food and Agriculture Organisation of the United Nations (FAO), Rome.

FAO (2012) *Coping with Water Scarcity: An Action Framework for Agriculture and Food Security*, Food and Agriculture Organisation, Rome.

Finley, C. (2009) 'The social construction of fishing, 1949', *Ecology and Society*, vol. 14, no. 1, article 6.

Fitzherbert, E. B., Struebig, M. J., Morel, A., Danielsen, F., Brühl, C. A., Donald, P. F. and Phalan, B. (2008) 'How will oil palm expansion affect biodiversity?', *Trends in Ecology & Evolution*, vol 23, no. 10, pp.538–545.

Folke, C. (2006) 'Resilience: the emergence of a perspective for social-ecological systems analysis', *Global Environmental Change*, vol. 16, pp.253–267.

Foster, S.S.D. and Perry, C.J. (2010) 'Improving groundwater resource accounting in irrigated areas: a prerequisite for promoting sustainable use', *Hydrogeology Journal*, vol. 18, pp.291–294.

Frederiksen, H.D. and Allen, R.G. (2011) 'A common basis for analysis, evaluation and comparison of offstream water uses', *Water International*, vol. 36, no. 3, pp.266–282.

Frederiksen, H.D., Allen, R.G., Burt, C.M. and Perry, C. (2012) 'Responses to Gleick *et al.* (2011), which was itself a response to Frederiksen and Allen (2011)', *Water International*, vol. 37, no. 2, pp.183–197.

Frisvold, G.B. and Deva, S. (2012) 'Farm size, irrigation practices, and conservation program participation in the US southwest', *Irrigation and Drainage*, vol. 61, no. 5, pp.569–582.

Galik, C.S. and Jackson, R.B. (2009) 'Risks to forest carbon offset projects in a changing climate', *Forest Ecology and Management*, vol. 257, pp.2209–2216.

Gandy, M. (1994) *Recycling and the Politics of Urban Waste*, Earthscan, London.

García-Amado, L.R., Pérez, M.R., Escutia, F.R., García, S.B. and Mejía, E.C. (2011) 'Efficiency of payments for environmental services: equity and additionality in a case study from a biosphere reserve in Chiapas, Mexico', *Ecological Economics*, vol. 70, pp.2361–2368.

Gavaris, S. (1996) 'Population stewardship rights: decentralized management through explicit accounting of the value of uncaught fish', *Canadian Journal of Fisheries and Aquatic Sciences* vol. 53, pp.1683–1691.

Geels, F.W. (2010) 'Ontologies, socio-technical transitions (to sustainability), and the multi-level perspective', *Research Policy*, vol. 39, pp.495–510.

Gibson, C., Ostrom, E. and Ahn, T.K. (2000) 'The concept of scale and the human dimensions of global change: a survey', *Ecological Economics,* vol. 32, pp.217–239.

Gleick, P.H., Christian-Smith, J. and Cooley, H. (2011) 'Water-use efficiency and productivity: rethinking the basin approach', *Water International*, vol. 36, no. 7, pp.784–798.

Godfrey, J. and Chalmers, K. (eds) (2012) *Water Accounting: International Approaches to Policy and Decision-Making*, Edward Elgar Publishing, Cheltenham.

Gordon, L.J., Finlayson, C.M. and Falkenmark, M. (2010) 'Managing water in agriculture for food production and other ecosystem services', *Agricultural Water Management*, vol. 97, no. 4, 512–519.

Graymore, M.L.M. and Wallis, A.M. (2010) 'Water savings or water efficiency? Water-use attitudes and behaviour in rural and regional areas', *International Journal of Sustainable Development and World Ecology*, vol. 17, no. 1, 84–93.

Greca, P.L., Barbarossa, L., Ignaccolo, M., Inturri, G. and Martinico, F. (2011) 'The density dilemma: A proposal for introducing smart growth principles in a sprawling settlement within Catania Metropolitan Area', *Cities*, vol. 28, no. 6, pp.527–535.

Grey, D. and Sadoff, C.W. (2007) 'Sink or swim? Water security for growth and development', *Water Policy*, vol. 9, pp.545–571.

Gupta, J. and van der Zaag, P. (2008) 'Interbasin water transfers and integrated water resources management: where engineering, science and politics interlock', *Physics and Chemistry of the Earth, Parts A/B/C*, vol. 33, pp.28–40.

Gutberlet, J. (2008) *Recovering Resources – Recycling Citizenship: Urban Poverty Reduction in Latin America*, Ashgate Publishing Ltd, Farnham.

Haie, N. and Keller, A.A. (2008) 'Effective dfficiency as a tool for sustainable water resources management', *Journal of the American Water Resources Association*, vol. 44, no. 4, pp.961–968.

Hannah, D. M., Demuth, S., van Lanen, H. A. J., Looser, U., Prudhomme, C., Rees, G., Stahl, K. and Tallaksen, L. M. (2011) 'Large-scale river flow archives: importance, current status and future needs', *Hydrological Processes*, vol. 25, pp.1191–1200.

Hardin, G. (1968) 'The tragedy of the commons', *Science*, vol. 162, pp.1243–1248.

Hays, S.P. (1959) *Conservation and the Gospel of Efficiency: The Progressive Conservation Movement, 1890–1920*. Harvard University Press, Cambridge, MA.

Heinmiller, B. T. (2009) 'Path dependency and collective action in common pool governance', *International Journal of the Commons*, vol. 3, no. 1, pp.131–147.

Heller, M.A. (1998) 'The tragedy of the anticommons: property in the transition from Marx to markets', *Harvard Law Review*, vol. 111, no. 3, pp.621–688.

Herring, H. (2006) 'Energy efficiency – a critical view', *Energy*, vol. 31, pp.10–20.

Hertwich, E.G. (2005) 'Consumption and the rebound effect: an industrial ecology perspective', *Journal of Industrial Ecology*, vol. 9, no. 1–2, pp.85–98.

Hess, C. (2008) 'Mapping the new commons', presented at 'Governing Shared Resources: Connecting Local Experience to Global Challenges', the Twelfth Biennial Conference of the International Association for the Study of the Commons, University of Gloucestershire, Cheltenham, July 14–18, 2008.

Hess, D. (2010) 'Sustainable consumption and the problem of resilience', *Sustainability: Science, Practice, and Policy*, vol. 6, no. 2, pp.26–37. Published online Sep 14, 2010. http://sspp.proquest.com/archives/vol6iss2/1001–005.hess.html

Hoekstra, A.K. (2008) *Water Neutral: Reducing and Offsetting the Impacts of Water Footprints*. Value of Water Research Report Series No. 28. Twente Water Centre, University of Twente, Twente.

Hoff, H. (2009) 'Global water resources and their management', *Current Opinion in Environmental Sustainability*, vol. 1, pp.141–147.

Homer-Dixon, T.F. (2001) *Environment, Scarcity, and Violence*, Princeton University Press, Princeton, NJ.

Hotta, Y., Elder, M., Mori, H. and Tanaka, M. (2008). 'Policy considerations for establishing an environmentally sound regional material flow in East Asia', *The Journal of Environment and Development*, vol. 17, no. 1, pp.26–50.

Huffaker, R., Whittlesey, N. and Hamilton, J.R. (2000) 'The role of prior appropriation in allocating water resources into the 21st century', *International Journal of Water Resources Development*, vol. 16, no. 2, pp.265–273.

Hufty, M. and Haakenstad, A. (2011) 'Reduced emissions for deforestation and degradation: a critical review', *Consilience: The Journal of Sustainable Development*, vol. 5, no. 1, pp.1–24.

ICID (1978) 'Describing irrigation efficiency and uniformity, International Commission on Irrigation and Drainage (ICID)', *Journal of the Irrigation and Drainage*, vol. 104, pp.35–41.

IRIN (2010) 'Syria: massive investment needed if Damascus to avert water crisis', IRIN news service, UN Office for the Coordination of Humanitarian Affairs. http://www.irinnews.org/report.aspx?reportid=61878. Accessed 14 October 2010.

Israelson, O.W. (1950) *Irrigation Principles and Practices*, John Wiley and Sons, New York.

Jensen, M.E. (1967) 'Evaluating irrigation efficiency', *Journal of the Irrigation and Drainage Division of the American Society of Civil Engineers (ASCE)*, vol. 93, no. 1, pp.83–98.

Jensen, M.E. (ed.) (1983) *Design and Operation of Farm Irrigation Systems*, American Society of Agricultural Engineers, St. Joseph, MI.

Jensen, M.E. (2007) 'Beyond irrigation efficiency', *Irrigation Science*, vol. 25, pp.233–245.

Jerneck, A. and Olsson, L. (2011) 'Breaking out of sustainability impasses: how to apply frame analysis, reframing and transition theory to global health challenges', *Environmental Innovation and Societal Transitions*, vol. 1, no. 2, pp.255–271.

Jollands, N. (2006) 'Concepts of efficiency in ecological economics: Sisyphus and the decision maker', *Ecological Economics*, vol. 56, no. 3, pp.359–372.

Jones, A., Pimbert, M. and Jiggins, J. (2011) *Virtuous Circles: Values, Systems and Sustainability*, International Institute for Environment and Development (IIED), London.

Karimi, P., Qureshi, A.S., Bahramloo, R. and Molden, D. (2012a) 'Reducing carbon emissions through improved irrigation and groundwater management: a case study from Iran', *Agricultural Water Management*, vol. 108, pp.52–60.

Karimi, P., Molden, D., Bastiaanssen, W. and Cai, X. (2012b) 'Water accounting to assess use and productivity of water: evolution of a concept and new frontiers', in J. Godfrey and K. Chalmers (eds), *Water Accounting: International Approaches to Policy and Decision-Making*, Edward Elgar Publishing, Cheltenham.

Karimov, A., Molden, D., Khamzina, T., Platonov, A. and Ivanov, Y. (2012) 'A water accounting procedure to determine the water savings potential of the Fergana Valley', *Agricultural Water Management*, vol. 108, pp.61–72.

Keller, A.A. and Keller, J. (1995) *Effective Efficiency: A Water Use Efficiency Concept for Allocating Freshwater Resources*, Center for Economic Policy Studies, Winrock International.

Keller, A.A., Keller, J. and Seckler, D. (1996) *Integrated Water Resource Systems: Theory and Policy Implications*. Research Report 3. International Water Management Institute (IWMI), Colombo, Sri Lanka.

Keys, P., Barron, J. and Lannerstad, M. (2012) *Releasing the Pressure: Water Resource Efficiencies and Gains for Ecosystem Services,* United Nations Environment Programme, Nairobi, Stockholm Environment Institute, Stockholm.

Knox, J.W., Kay, M.G. and Weatherhead E.K. (2012) 'Water regulation, crop production, and agricultural water management – understanding farmer perspectives on irrigation efficiency', *Agricultural Water Management*, vol. 108, pp.3–8.

Kraatz, D.B. and Mahajan, I.K. (1975) Small Hydraulic Structures. FAO Irrigation and Drainage Paper Nos. 26/1 and 26/2, Food and Agriculture Organisation, Rome, Italy.

Lal, R. (2008) 'Carbon sequestration', *Philosophical Transactions of the Royal Society B: Biological Sciences*, vol. 363, no. 1492, pp.815–830.

Lang, T. (2003) 'Food industrialisation and food power: implications for food governance', *Development Policy Review*, vol. 21, no. 5–6, pp.555–568.

Lankford, B. A. (1992) ' The use of measured water flows in furrow irrigation management – a case study in Swaziland', *Irrigation and Drainage Systems*, vol. 6, pp.113–128.

Lankford, B.A. (2004) 'Resource-centred thinking in river basins: should we revoke the crop water approach to irrigation planning?', *Agricultural Water Management*, vol. 68, no. 1, pp.33–46.

Lankford, B.A. (2006) 'Localising irrigation efficiency', *Irrigation and Drainage Systems*, vol. 55, pp.345–362.

Lankford, B.A. (2008) 'Book review "Multi-stakeholder platforms for integrated water management edited by Jeroen Warner, Ashgate, 2007"', *Journal of Environmental Planning and Management*, vol. 51, no. 2, pp.321–322.

Lankford, B.A. (2011) 'Responding to water scarcity – beyond the volumetric', in L. Mehta, (ed.) *The Limits to Scarcity: Contesting the Politics of Allocation*, Earthscan, London.

Lankford, B.A. (2012a) 'Preface: towards a political ecology of irrigation efficiency and productivity', *Journal of Agricultural Water Management*, vol. 108, pp.1–2.

Lankford, B.A. (2012b) 'Fictions, fractions, factorials, fractures and fractals; on the framing of irrigation efficiency', *Agricultural Water Management*, vol. 108, pp.27–38.

Lankford B.A. and Beale, T. (2007) 'Equilibrium and non-equilibrium theories of sustainable water resources management: dynamic river basin and irrigation behaviour in Tanzania', *Global Environmental Change*, vol. 17, no. 2, pp.168–180.

Lankford, B.A. and Gowing, J. (1996a) 'The impact of design approximations on the operational performance of an irrigation scheme: a case study in Malaysia', *Irrigation and Drainage Systems*, vol. 10, pp.193–205.

Lankford, B.A. and Gowing, J. (1996b) 'Understanding water supply control in canal irrigation systems', in P. Howsan and R. Carter (eds) *Water Policy: Allocation and Management in Practice*, E. & F.N. Spon, London.

Lankford, B.A. and Mwaruvanda, W. (2007) 'A legal-infrastructural framework for catchment apportionment', in B. Van Koppen, M. Giordan, and J. Butterworth (eds), *Community-based Water Law and Water Resource Management Reform in Developing Countries*, Comprehensive Assessment of Water Management in Agriculture Series, CABI Publishing, Wallingford.

Lankford, B.A. and Watson, D. (2007) 'Metaphor in natural resource gaming; insights from the River Basin Game', *Simulation & Gaming*, vol. 38, no. 3, pp.421–442.

Lankford, B.A., Tumbo, S. and Rajabu, K. (2009) 'Water competition, variability and river basin governance: a critical analysis of the Great Ruaha River, Tanzania', in F. Molle and P. Wester (eds) *River Basin Development in Perspective*, CABI, Wallingford.

Larson, W.L. and Richter, B. (2009) 'The role of "water offsets" in water stewardship certification', *Journal of the American Water Works Association*, pp.40–44.

Law, B.E. and Harmon, M.E. (2011) 'Forest sector carbon management, measurement and verification, and discussion of policy related to climate change', *Carbon Management*, vol. 2, no. 1, pp.73–84.

Lawrence, M. (1997) 'Heartlands or neglected geographies? Liminality, power, and the hyperreal rural', *Journal of Rural Studies*, vol. 13, no. 1, pp.1–17.

Leach, M., Fairhead, J. and Fraser, J. (2012) 'Green grabs and biochar: revaluing African soils and farming in the new carbon economy', *The Journal of Peasant Studies*, vol. 39, no. 2, pp.285–307.

Leach, M., Scoones, I. and Stirling, A. (2010) *Dynamic Sustainabilities: Technology, Environment, Social Justice*, Earthscan, London.

Lehmann, J. (2007) 'A handful of carbon', *Nature*, vol. 447, no. 7141, pp.143–144.

Leibenstein, H. (1966) 'Allocative efficiency vs. "X-efficiency"', *The American Economic Review*, vol. 56, no. 3, pp.392–415.

Lenton, R. and Muller M., (2009) *Integrated Water Resources Management in Practice: Better Water Management for Development*, Earthscan, London.

Lewis, D. (2009) 'International development and the "perpetual present": anthropological approaches to the re-historicization of policy', *European Journal of Development Research*, vol. 21, no. 1, pp.32–46.

Li, T-M. (2007) *The Will to Improve: Governmentality, Development, and the Practice of Politics*, Duke University Press Books, Durham, NC.

Linton, J. (2008) 'Is the hydrologic cycle sustainable? A historical–geographical critique of a modern concept', *Annals of the Association of American Geographers*, vol. 98, no. 3, pp.630–649.

Lopez-Gunn, E., Zorrilla, P., Prieto, F. and Llamas, R. (2012) 'Lost in translation? Water efficiency in Spanish agriculture', *Agricultural Water Management*, vol. 108, pp.83–95.

Low, N. and Gleeson, B. (1998) *Justice, Society, and Nature: An Exploration of Political Ecology*, Routledge, London.

Lundqvist, J. (2009) 'Unpredictable and significant variability of rainfall: carryover stocks of water and food necessary', *Review of Environmental Science and Bio/Technology*, vol. 8, no. 3, pp.219–223.

McCartney, M.P., Lankford, B.A. and Mahoo, H.F. (2007) *Agricultural Water Management in a Water Stressed Catchment: Lessons from the RIPARWIN Project*. Research Report 116. International Water Management Institute, Colombo, Sri Lanka.

Machibya, M. (2003) 'Challenging established concepts of irrigation efficiency in a water scarce river basin: a case study of the Usangu Basin, Tanzania', PhD thesis, University of East Anglia.

Mainuddin, M., Kirby, M. and Hoanh, C.T. (2012) 'Water productivity responses and adaptation to climate change in the lower Mekong basin', *Water International*, vol. 37, no. 1, pp.53–74.

Makoff, R. (2011) 'Confronting climate crisis: a framework for understanding the criteria for addressing dangerous climate change', PhD thesis, University of East Anglia.

Manson, S.M. (2001) 'Simplifying complexity: a review of complexity theory', *Geoforum*, vol. 32, no. 3, pp.405–414.

Manson, S.M. (2008) 'Does scale exist? An epistemological scale continuum for complex human-environment systems', *Geoforum*, vol. 39, no. 2, pp.776–788.

Mateos, L., Lozano, D., Ould Baghil, A.B., Diallo, O.A., Gómez-Macpherson, H., Comas, J. and Connor, D. (2010) 'Irrigation performance before and after rehabilitation of a representative, small irrigation scheme besides the Senegal River, Mauritania', *Agricultural Water Management*, vol. 97, no. 6, pp.901–909.

Matthews, J.H., Wickel, B.A. and Freeman, S. (2011) 'Converging currents in climate-relevant conservation: water, infrastructure, and institutions', *PLoS Biology*, vol. 9, no. 9.

McCay, B.J. (1996) 'Common and private concerns', in S. Hanna, C. Folke and K-G Mäler (eds), *Rights to Nature: Ecological, Economic, Cultural, and Political Principles of Institutions for the Invironment*, Island Press, Washington, DC.

MEA (2005) *Ecosystems and Human Well-being: Synthesis. Millennium Ecosystem Assessment*, Island Press, Washington, DC.

Medellín-Azuara, J., Howitt, R.E. and Harou, J.J. (2012) 'Predicting farmer responses to water pricing, rationing and subsidies assuming profit maximizing investment in irrigation technology', *Agricultural Water Management*, vol. 108, pp.73–82.

Mediterranean Water Scarcity and Drought Working Group (2007) Mediterranean Water Scarcity and Drought. Technical report on Water Scarcity and Drought Management in the Mediterranean and the Water Framework Directive. Technical Report – 009– 2007.

Molden, D. and Sakthivadivel, R. (2006) 'Water accounting to assess use and productivity of water', *International Journal of Water Resources Development*, vol. 15, pp.55–71.

Molden, D., Oweis, T., Steduto, P., Bindraban, P., Hanjra, M.A. and Kijne, J. (2010) 'Improving agricultural water productivity: between optimism and caution', *Agricultural Water Management*, vol. 97, pp.528–535.

Molle, F. and Berkoff, J. (2009) 'Cities vs. agriculture: A review of intersectoral water re-allocation', *Natural Resources Forum*, vol. 33, no. 1, pp.6–18.

Mooney, H., Larigauderie, A., Cesario, M., Elmquist, T., Hoegh-Guldberg, O., Lavorel, S., Mace, G.M., Palmer, M., Scholes, R. and Yahara, T. (2009) 'Biodiversity, climate change, and ecosystem services', *Current Opinion in Environmental Sustainability*, vol. 1, no. 1, pp.46–54.

Mueller, K., Cao, L., Caldeira, K. and Jain, A. (2004) 'Differing methods of accounting ocean carbon sequestration efficiency', *Journal of Geophysical Research (Oceans)*, vol. 109, C12018.

Muradian, R. and Rival, L. (2013) *Governing the Provision of Ecosystem Services*, Studies in Ecological Economics, Springer Science and Business Media, Dordrecht.

Murray-Rust, D.H. and Snellen, W.B. (1993) *Irrigation System Performance Assessment and Diagnosis*, International Irrigation Management Institute, Sri Lanka.

Narayanamoorthy, A. (2004) 'Impact assessment of drip irrigation in India: the case of sugarcane', *Development Policy Review*, vol. 22, no. 4, pp.443–462.

Nelson, E., Polasky, S., Lewis, D.J., Plantinga, A.J., Lonsdorf, E., White, D., Bael, D. and Lawler, J.J. (2008) 'Efficiency of incentives to jointly increase carbon sequestration and species conservation on a landscape', *Proceedings National Academy Science USA*, vol. 105, no. 28, pp.9471–9476.

Neuman, J.C. (1998) 'Beneficial use, waste, and forfeiture: the inefficient search for efficiency in western water use', *Environmental Law*, vol. 28, pp.919–996.

Norgaard, R.B. (2010) 'Ecosystem services: from eye-opening metaphor to complexity blinder', *Ecological Economics*, vol. 69, no. 6, pp.1219–1227.

Norris, J. (2011) 'Montana v. Wyoming: Is water conservation drowning the Yellowstone River compact', *University of Denver Water Law Review*, vol. 15, 189.

Odum, T.H. (1996) *Environmental Accounting: Emergy and Environmental Making*, John Wiley and Sons, Toronto.

OECD (2011) *OECD Green Growth Studies: Energy*, The Organisation for Economic Co-operation and Development (OECD), Paris.

OED (2012) *Oxford English Dictionary* online edition. http://www.oed.com/

Ollivier, H. (2012) 'Growth, deforestation and the efficiency of the REDD mechanism', *Journal of Environmental Economics and Management*, vol. 64, no. 3, pp.312–327.

Olschewski, R. and Klein, A. (2011) 'Ecosystem services between sustainability and efficiency', *Sustainability: Science, Practice, & Policy*, vol. 7, no. 1, pp.69–73.

Ostendorf, B. (2011) 'Overview: spatial information and indicators for sustainable management of natural resources', *Ecological Indicators*, vol. 11, no. 1, pp.97–102.

Ostrom, E. (1990) *Governing the Commons: The Evolution of Institutions for Collective Action*, Cambridge University Press, Cambridge.

Ostrom, E., Burger, J., Field, C.B., Norgaard, R.B. and Policansky, D. (1999) 'Revisiting the commons: local lessons, global challenges', *Science*, vol. 284, no. 5412, pp.278–282.

Ott, C., Windsperger, A., Windsperger, B. and Hummel, M. (2011) 'Optimizing resource efficiency and carbon intensity in the wood processing sector in Austria', in R. Bleischwitz, P.J.J. Welfens and Z. Zhang (eds) *International Economics of Resource Efficiency: Eco-Innovation Policies for a Green Economy*, Springer, Heidelberg.

Paquin, R. and Howard-Grenville, J. (2009) 'Facilitating regional industrial symbiosis: network growth in the UK's national industrial symbiosis programme', in F. Boons and J. Howard-Grenville (eds) *The Social Embeddedness of Industrial Ecology*, Edward Elgar, Cheltenham.

Payero, J.O., Tarkalson, D.D., Irmak, S., Davison, D. and Petersen, J.L. (2009) 'Effect of timing of a deficit-irrigation allocation on corn evapotranspiration, yield, water use efficiency and dry mass', *Agricultural Water Management*, vol. 96, pp.1387–1397.

Pearce, F. (2007) *When The Rivers Run Dry: What Happens When Our Water Runs Out?* Eden Project Books, UK.

Pease, M. (2012) 'Water transfer laws and policies: tough questions and institutional reform for the Western United States', *Journal of Natural Resources Policy Research*, vol. 4, no. 2, pp.103–119.

Pereira, L.S, Cordery, I. and Iacovides, I. (2012) 'Improved indicators of water use performance and productivity for sustainable water conservation and saving', *Agricultural Water Management*, vol. 108, pp.39–51.

Perry, C. (2011) 'Accounting for water use: terminology and implications for saving water and increasing production', *Agricultural Water Management*, vol. 98, no. 12, pp.1840–1846.

Perry, C.J. (2007) 'Efficient irrigation; inefficient communication; flawed recommendations', *Irrigation and Drainage*, vol. 56, pp.367–378.

Phillips, J.D. (2008) 'Goal functions in ecosystem and biosphere evolution', *Progress in Physical Geography*, vol. 32 no. 1, pp.51–64.

PI (2010) *California's Next Million Acre-feet: Saving Water, Energy and Money*, Pacific Institute, Oakland, CA.

Pittock, J. and Lankford, B.A. (2010) 'Environmental water requirements: demand management in an era of water scarcity', *Journal of Integrative Environmental Sciences*, vol. 7, no. 1, pp.1–19.

Playán, E. and Mateos, L. (2006) 'Modernization and optimization of irrigation systems to increase water productivity', *Agricultural Water Management*, vol. 80, pp.100–116.

Plusquellec, H., Burt, C. and Wolter, H.W. (1994) *Modern Water Control in Irrigation. Concepts, Issues, and Applications*. World Bank Technical Paper No. 246. Irrigation and Drainage Series. World Bank, Washington, DC.

Polimeni, J.M., Mayumi, K., Giampietro, M. and Alcott, B. (2008) *The Jevons Paradox and the Myth of Resource Efficiency Improvements*, Earthscan, London.

Poulton, C., Kydd, J. and Dorward, A. (2006) 'Overcoming market constraints on pro-poor agricultural growth in Sub-Saharan Africa', *Development Policy Review*, vol. 24, no. 3, pp.243–277.

Princen, T. (2003) 'Principles for sustainability: from cooperation and efficiency to sufficiency', *Global Environmental Politics*, vol. 3, no. 1, pp.33–50.

Princen, T. (2005) *The Logic of Sufficiency*, MIT, Cambridge, MA.

Qadir, M., Wichelns, D., Raschid-Sally, L., McCornick, P.G., Drechsel, P., Bahri, A. and Minhas, P.S. (2010) 'The challenges of wastewater irrigation in developing countries', *Agricultural Water Management*, vol. 97, pp.561–568.

Rajabu, K.R.M. (2007) 'Use and impacts of the river basin game in implementing integrated water resources management in Mkoji sub-catchment in Tanzania', *Agricultural Water management*, vol. 94, no. 1–3, pp.63–72.

Rammel, C., Stagl, S. and Wilfing, H. (2007) 'Managing complex adaptive systems – a co-evolutionary perspective on natural resource management', *Ecological Economics*, vol. 63, no. 1, pp.9–21.

Rappocciolo, F. (2012) *Spreading the System of Rice Intensification across East and Southern Africa*, Seeds of Innovation. IFAD, www.ifad.org/operations/projects/regions/pf/seeds/6.htm

Raymond, E. (2000) 'The magic cauldron', first published 1999 in ALS '99: Proceedings of the Third Annual Conference on Atlanta Linux Showcase – vol. 3, USENIX Association, Berkeley, CA.

Reijnders, L. (1998) 'The factor "X" debate: Setting targets for eco-efficiency', *Journal of Industrial Ecology*, vol. 2, no. 1, pp.13–22.

Renault, D., Facon, T. and Wahaj, R. (2007) *Modernizing Irrigation Management – The MASSCOTE Approach*, Food and Agriculture Organization (FAO), Rome.

Replogle, J.A. (1989) 'Obstacles to flow control and measurement in irrigation practice', in J.R. Rydzewski, and C.F Ward (eds) *Irrigation; Theory and Practice*. Proceedings of the International Conference held at the University of Southampton, 12–15 September 1989, London: Pentech Press.

Rice, J. (n.d.) *Carbon Stock Conditions and How Climate and Disturbance May Influence Carbon dynamics on the Shoshone National Forest, Wyoming*, Shoshone National Forest Plan. www.fs.usda.gov/Internet/FSE_DOCUMENTS/stelprdb5379186.pdf

Robbins, M.W. (2011) *Crops and Carbon: Paying Farmers to Combat Climate Change*, Earthscan, London.

Rocheleau, D.E. (2008) 'Political ecology in the key of policy: from chains of explanation to webs of relation', *Geoforum*, vol. 39, pp.716–727.

Rockström, J. *et al.*. 2009. 'Planetary boundaries: exploring the safe operating space for humanity', *Ecology and Society*, vol. 14, no. 2, p.32.

Rogers, P. and Leal, S. (2010) *Running Out of Water: The Looming Crisis and Solutions to Conserve Our Most Precious Resource*, Palgrave Macmillan, New York.

Rogers, P., de Silva, R. and Bhatia, R. (2002) 'Water is an economic good: how to use prices to promote equity, efficiency, and sustainability', *Water Policy*, vol. 4, no. 1, pp.1–17.

Rothausen, S.G.S.A. and Conway, D. (2011) 'Greenhouse-gas emissions from energy use in the water sector', *Nature Climate Change*, vol. 1, pp.210–219.

Runhaar, H., Runhaar, P.R. and Oegema, T. (2010) 'Food for thought: conditions for discourse reflection in the light of environmental assessment', *Environmental Impact Assessment Review*, vol. 30, no. 6, pp.339–346.

Ruth, M., Kalnay, E., Zeng, N., Rachel, S., Franklin, R.S., Rivas, J. and Miralles-Wilhelm, F. (2011) 'Sustainable prosperity and societal transitions: long-term modeling for anticipatory management', *Environmental Innovation and Societal Transitions*, vol. 1, no. 1, pp.160–165.

Ruzzenenti, F. and Basosi, R. (2008) 'The role of the power/efficiency misconception in the rebound effect's size debate: does efficiency actually lead to a power enhancement?', *Energy Policy*, vol. 36, no. 9, pp.3626–3632.

Ryan, M.G., Harmon, M.E., Birdsey, R.A., Giardina, C.P., Heath, L.S., Houghton, R.A., Jackson, R.B., McKinley, D.C., Morrison, J.F., Murray, B.C., Pataki, D.E. and Skog, K.E. (2010) *A Synthesis of the Science on Forests and Carbon for U.S. Forests*. Issues in Ecology Report 13. Ecological Society of America, Washington, DC.

Sagardoy, J.A. and Bottrall, A. (1982) *Organisation, Operation and Maintenance of Irrigation Schemes*. FAO Irrigation and Drainage Paper No. 40, Food and Agriculture Organisation, Rome.

Salman, S.M.A. (2010) 'Downstream riparians can also harm upstream riparians: the concept of foreclosure of future uses', *Water International*, vol. 35, no. 4, pp.350–364.

Sax, J.L. (1990) 'The constitution, property rights and the future of water law', *University of Colorado Law Review*, vol. 61, pp.257–282.

Schneider, M. and Somers, M. (2006) 'Organizations as complex adaptive systems: implications of Complexity Theory for leadership research', *Leadership Quarterly*, vol. 17, no. 4, pp.351–365.

Schroeder, C.H. (2000) 'Third way environmentalism,' *University of Kansas Law Review*, vol. 48, pp.1–48.

Scoones, I. (1999) 'New ecology and the social sciences: what prospects for a fruitful engagement?', *Annual Review of Anthropology*, vol. 28, pp.479–507.

Scott, C.A. and Raschid-Sally, L. (2012) 'The global commodification of wastewater', *Water International*, vol. 37, no. 2, pp.147–155.

Seckler, D.W. (1996) *The New Era of Water Resources Management: From 'Dry' to 'Wet' Water Savings*, International Irrigation Management Institute, Colombo.

Seckler, D.W., Molden, D. and Sakthivadivel, R. (2003) 'The concept of efficiency in water resources management and policy', in J.W. Kijne, R. Barker and D. Molden (eds) *Water Productivity in Agriculture: Limits and Opportunities for Improvement*, CABI Publishing, Wallingford.

SEI (2011) *Background Paper for the Bonn 2011 Nexus Conference: The Water, Energy and Food Security Nexus*, Stockholm Environment Institute, Stockholm.

Shupe, S.J. (1982) 'Waste in western water law: a blueprint for change', *Oregon Law Review*, vol. 61, pp.483–499.

Sikor, T. (ed.) (2013) *The Justices and Injustices of Ecosystem Services*, Earthscan, London.

Singleton, S. (1999) 'Commons problems, collective action and efficiency: past and present institutions of governance in Pacific Northwest salmon fisheries', *Journal of Theoretical Politics*, vol. 11, no. 3, pp.367–391.

Sinn, H.-W. (2012) *The Green Paradox: A Supply-Side Approach to Global Warming*, MIT Press, Cambridge, MA.

Skaggs, R., Samani, Z., Bawazir, A.S. and Bleiweiss, M. (2011) 'The convergence of water rights, structural change, technology, and hydrology: a case study of new Mexico's Lower Rio Grande', *Natural Resources Journal*, vol. 51, no. 1, pp.95–119.

Skyttner, L. (2006) *General Systems Theory: Problems, Perspectives, Practice*, 2nd edition, World Scientific Publishing Co., London.

Small, L.E. and Carruthers, I. (1991) *Farmer-financed Irrigation: The Economics of Reform*. Wye Studies in Agricultural and Rural Development. Cambridge University Press, Cambridge.

Smith, H. (2008) 'Governing water: the semicommons of fluid property rights', *Arizona Law Review*, vol. 50, no. 2, pp.445–478.

Snaddon, C.D., Wishart, M.J. and Davies, B.R. (1998) 'Some implications of inter-basin water transfers for river ecosystem functioning and water resources management in southern Africa', *Aquatic Ecosystem Health and Management*, vol. 1, pp.159–182.

Snap, S. and Pound, B. (eds) (2008) *Agricultural Systems: Agroecology and Rural Innovation for Development*, Academic Press, London.

Sneddon, C. (2007) 'Nature's materiality and the circuitous paths of accumulation: dispossession of freshwater fisheries in Cambodia', *Antipode*, vol. 39, no. 1, pp.167–93.

Socolow, R.H., Andrews, C., Berkhout, F. and Thomas, V. (1996) *Industrial Ecology and Global Change*, Cambridge University Press, Cambridge.

Solanes, M. and Gonzalez-Villarreal, F. (1999) *The Dublin Principles for Water as Reflected in a Comparative Assessment of Institutional and Legal Arrangements for Integrated Water Resources Management*, Global Water Partnership (GWP), Stockholm.

Solomon, S. (2010) *Water: The Epic Struggle for Wealth, Power, and Civilization*, Harper Perennial, New York.

Solomon, K.H. and Burt, C.M. (1999) 'Irrigation sagacity: a measure of prudent water use', *Irrigation Science*, vol. 18, pp.135–140.

Sorrell, S. (2009) 'Jevons' paradox revisited: the evidence for backfire from improved energy efficiency', *Energy Policy*, vol. 37, no. 4, pp.1456–1469.

Sorrell, S. and Dimitropoulos, J. (2008) 'The rebound effect: microeconomic definitions, limitations and extensions,' *Ecological Economics*, vol. 65, no. 3, pp.636–649.

Souza, E.L. and Ghisi, E. (2012) 'Potable water savings by using rainwater for non-potable uses in houses', *Water*, vol. 4, pp.607–628.

Spronk, S. (2010) 'Water and sanitation utilities in the global south: re-centering the Debate on "Efficiency"', *Review of Radical Political Economics*, vol. 42, no. 2, pp.156–174.

Storper, M. (2011) 'Justice, efficiency and economic geography: should places help one another to develop?', *European Urban and Regional Studies*, vol. 18, no. 1, pp.3–21.

Sullivan, S. (2013) 'After the green rush? Biodiversity offsets, uranium power and the "calculus of casualties" in greening growth', *Human Geography*, vol. 6, no. 1, pp.80–101.

Svendsen, M. (2006) *Irrigation and River Basin Management: Options for Governance and Institutions*, CABI Publishing., Wallingford.

Tainter, J.A. (2011) 'Energy, complexity, and sustainability: a historical perspective', *Environmental Innovation and Societal Transitions*, vol. 1, no. 1, pp.89–95.

Tarlock, D. and Wouters, P. (2007) 'Are shared benefits of international waters an equitable appportionment?', *Colorado Journal of International Environmental Law and Policy*, vol. 18, pp.523–536.

The Economist (2010) 'For want of a drink: A special report on water', *The Economist*, London.

Thompson, M.C., Baruah, M. and Carr, E.R. (2011) 'Seeing REDD+ as a project of environmental governance', *Environmental Science and Policy*, vol. 14, no. 2, pp.100–110.

Tortajada, C. (2006) 'Water management in Singapore', *International Journal of Water Resources Development*, vol. 22, pp.227–240.

Towler, R.W. (1975) 'Forestry in national parks', *Quarterly Journal of Forestry*, vol. LXIX, no. 3, pp.129–136.

Trawick, P.B. (2001) 'Successfully governing the commons: principles of social organization in an Andean irrigation system', *Human Ecology*, vol. 29, no. 1, pp.1–25.

Underdal, A. (2010) 'Complexity and challenges of long-term environmental governance', *Global Environmental Change*, vol. 20, no. 3, pp.386–393.

UNEP (2012) *Measuring Water use in a green economy, A Report of the Working Group on Water Efficiency to The International Resource Panel*. J. McGlade, B. Werner, M. Young, M. Matlock, D. Jefferies, G. Sonnemann, M. Aldaya, S. Pfister, M. Berger, C. Farell, K. Hyde, M.

Wackernagel, A. Hoekstra, R. Mathews, J. Liu, E. Ercin, J.L. Weber, A. Alfieri, R. Martinez-Lagunes, B. Edens, P. Schulte, S. von Wirén-Lehrand and D. Gee, United Nations Environment Programme (UNEP), New York.

UNSD (2007) *United Nations System for Environmental-Economic Accounting for Water (UNSEEAW)*, United Nations Statistics Division, New York.

van der Kooij S., Zwarteveen, M., Boesveld, H. and Kuper, M. (forthcoming) 'The efficiency of drip irrigation unpacked', *Agricultural Water Management*.

van der Ploeg, F. and Withagen, C. (2012) 'Is there really a green paradox?' *Journal of Environmental Economics and Management*, vol. 64, no. 3, pp.342–363.

van Gennep, A. (1909) *The Rites of Passage*, issued in English by Routledge and Kegan Paul, London, 1960, and University of Chicago Press, 1961.

van Halsema, G.E. and Vincent, L. (2012) 'Efficiency and productivity terms for water management: a matter of contextual relativism versus general absolutism', *Agricultural Water Management*, 108, pp.9–15.

Varela-Ortega, C. and Sagardoy, J.A. (2003) 'Irrigation water policies in Syria: current developments and future options', in C. Fiorillo and J. Vercueil (eds), *Syrian Agriculture at the Crossroads*, FAO Agricultural Policy and Economic Development Series No. 8. Food and Agriculture Organization of the United Nations, Rome.

Wagner, J.R. (2012) 'Water and the commons imaginary', *Current Anthropology*, vol. 53, no. 5, 617–641.

Wagner, M. and Wellmer, F.-W. (2009) 'A hierarchy of natural resources with respect to sustainable development – a basis for a natural resources efficiency indicator', in J.P. Richards (ed.), *Mining, Society and a Sustainable World*, Springer, Heidelberg.

Walker, G. (2012) *Environmental Justice: Concepts, Evidence and Politics*. Routledge, London.

Walker, B., Holling, C.S., Carpenter, S.R. and Kinzig, A. (2004) 'Resilience, adaptability and transformability in social–ecological systems', *Ecology and Society*, vol. 9, no. 2, article 5.

Ward, F.A. and Pulido-Velázquez, M. (2008) 'Water conservation in irrigation can increase water use', *Proceedings of the National Academy of Sciences*, vol. 105, no. 47, pp.18215–18220.

Warner, J. (ed.) (2007) *Multi-stakeholder Platforms for Integrated Water Management*, Ashgate, Aldershot.

Warner, R. (2010) 'Ecological modernization theory: towards a critical ecopolitics of change?', *Environmental Politics*, vol. 19, no. 4, pp.538–556.

WBCSD (2000) *Eco-Efficiency: Creating More Value with Less Impact*, World Business Council for Sustainable Development (WBCSD). Available online. http://www.wbcsd.org/web/publications/eco_efficiency_creating_more_value.pdf

Wichelns, D. (2011) 'Assessing water footprints will not be helpful in improving water management or ensuring food security', *International Journal of Water Resources Development*, vol. 27, no. 3, pp.607–619.

Wichelns, D. and Drechsel, P. (2011) 'Meeting the challenge of wastewater irrigation: economics, finance, business opportunities and methodological constraints', *Water International*, vol. 36, no. 4, pp.415–419.

Wildavsky, A. (1966) 'The political economy of efficiency: cost-benefit analysis, systems analysis, and program budgeting', *Public Administration Review*, vol. 26, no. 4, pp.292–310.

Willardson, L.S., Allen, R.G. and Frederiksen, H. (1994) 'Eliminating irrigation efficiencies', USCID Thirteenth Technical Conference, Denver, Colorado, 19–22 October 1994.

Wilson, B. (1990) *Systems: Concepts, Methodologies and Applications*, second edition, John Wiley and Sons, Chichester.

Wisser, D., Frolking, S., Douglas, E.M., Fekete, B.M., Schumann, A.H. and Vörösmarty, C.J. (2010) 'The significance of local water resources captured in small reservoirs for crop production – a global-scale analysis', *Journal of Hydrology*, vol. 384, nos. 3–4, pp.264–275.

Woodhouse, E.J. and Nieusma, D. (2001) 'Democratic expertise: integrating knowledge, power, and participation' in M. Hisschemöller, R. Hoppe, W. N. Dunn and J. Ravetz (eds) *Knowledge Use and Political Choice in Environmental Policy Analysis* Transaction, New Brunswick, NJ.

World Commission on Dams (WCD) (2000) *Dams and Development: A New Framework for Decision-making. Final Report*, Earthscan Publications, London.

World Water Assessment Programme (2009) *The United Nations World Water Development Report 3: Water in a Changing World*, UNESCO Publishing, Paris and Earthscan, London.

Yanlk, L.K. (2011) 'Constructing Turkish "exceptionalism": discourses of liminality and hybridity in post-cold war Turkish foreign policy', *Political Geography*, vol. 30, no. 2, pp.80–89.

York, R. and Rosa, E.A. (2003) 'Key challenges to ecological modernization theory: institutional efficacy, case study evidence, units of analysis, and the pace of eco-efficiency', *Organization and Environment*, vol. 16, no. 3, pp.273–288.

Young, M. (2012) 'Towards a generic framework for the abstraction and utilisation of water in England and Wales'. Visiting Fellowship Report for DEFRA (Department for Environment, Food and Rural Affairs), University of Adelaide, The Environment Institute, University of Adelaide, University College London.

Zeitoun, M. and Warner, J. (2006) 'Hydro-hegemony – a framework for analysis of trans-boundary water conflicts', *Water Policy*, vol. 8, pp.435–460.

Zwarteveen, M. (2008) 'Men, masculinities and water powers in irrigation', *Water Alternatives*, vol. 1, no. 1, pp.111–130.

Zwarteveen, M. (2010) 'A masculine water world. The politics of gender and identity in irrigation expert thinking', in R. Boelens, D. Getches and A. Guevara (eds), *Out of the Mainstream: Water Rights, Politics and Identity*, Earthscan, London.

Author index

Subject index